Studying the Novel

An Introduction

Third edition

JEREMY HAWTHORN

Professor of Modern British Literature,
Norwegian University of Science and
Technology, Trondheim

A member of the Hodder Headline Group
LONDON • NEW YORK • SYDNEY • AUCKLAND

First published in Great Britain in 1985
Second edition published in 1992
Third edition published in 1997 by
Arnold, a member of the Hodder Headline Group
338 Euston Road, London NW1 3BH
175 Fifth Avenue, New York, NY 10010

Distributed exclusively in the USA by
St Martin's Press Inc.
175 Fifth Avenue, New York, NY 10010

British Library Cataloguing in Publication Data
A catalogue entry for this book is available from the British Library

Library of Congress Cataloging-in-Publication Data
Hawthorn, Jeremy.
    Studying the novel : an introduction / Jeremy Hawthorn. — 3rd ed.
        p.   cm.
    Includes bibliographical references and index.
    ISBN 0-340-69220-0 (pbk. : alk. paper)
    1. Fiction—History and criticism.   2. Criticism.   I. Title.
PN3335.H34   1997
808.3—dc21   97-6752
                                                                        CIP

ISBN 0 340 69220 0

Typeset by the author
Printed and bound in Great Britain by
J W Arrowsmith Ltd, Bristol

THE UNIVERSITY OF WINCHESTER

Contents

Introduction to the Third Edition

I finished writing the first edition of this book in April 1985 – nearly twelve years ago. At that time I was teaching a first-year course on the novel, and the introduction to the genre that I had used as a set text had gone out of print. When I mentioned this to Christopher Wheeler – my editor then and now at Arnold – he suggested that I fill the gap by writing a replacement text myself.

Although I was not fully aware of it then, it was a propitious time to write such a book. The achievements of structuralist narratology had added to and refined the analytical tools available to literary critics, so that there was much new material waiting to be gathered in an introductory textbook. Terms such as 'anachrony', 'analepsis', 'ellipsis', 'frequency', 'perspective and voice' were at that time more likely to be found in advanced monographs than in introductory textbooks, even though their usefulness seemed to me to be undeniable. Thus in spite of my Anglo-Saxon pragmatism, my hostility to the ahistoricism of structuralism, and my middle-aged resistance to changing the way that I myself talked about works of fiction, I saw that the achievements of structuralist narratology ought to be made generally available to students of the novel.

At the same time, that period of unquestioning assent to Ian Watt's account of 'the rise of the novel' was coming to an end. This is not to say that Watt's account was (or is) rejected *in toto*, but that extensive research work on the prose fiction of the sixteenth, seventeenth, and eighteenth centuries was allowing scholars to build up a more nuanced and complex picture of the early days of the modern novel, a process which continues today.

Moreover, the 'critical theory explosion' had taken place, and the end of New Critical hegemony (which had, anyway, never been quite so complete among critics of the novel as it had been among critics of drama and poetry) meant that novels and short stories could be seen in a range of different explanatory contexts and in terms of alternative interpretative approaches and analytical strategies which balanced and sharpened one another. In particular, the work of Mikhail Bakhtin – newly available to English-speaking readers in the 1980s – has been very important to me in the writing of successive editions of this book.

Finally, the work of contemporary novelists and short story writers was such as to challenge static and conventional views of the genre – as it always has been. Introductions to the novel will always have to be regularly revised and rewritten because 'the novel' itself never stands still: it is a genre in a state of permanent revolution.

In this third edition I have tried to make my comments a little less Eurocentric – referring to some non-European novelists and talking about the novel as an international literary genre. I have also attempted to respond to some of the ideas and arguments advanced by postcolonialist critics and to indicate that even the domestic European novel (and especially the English novel) bears the marks of Europe's chequered history and relations with the rest of the world. And in response to a range of interesting work by philosophers and aestheticians on the concept of fiction I have tried to indicate some of the scope, complexity, and paradoxical nature of this familiar term.

As in previous editions I have restricted referencing to a minimum, omitting sources for standard critical and fictional works. When I mention novels and short stories for the first time I provide the date of initial publication.

Jeremy Hawthorn

Trondheim, Norway
January 1997

1 Introductory Definitions

According to the *Oxford English Dictionary* a novel is 'a fictitious prose narrative or tale of considerable length (now usually one long enough to fill one or more volumes) in which characters and actions representative of the real life of past or present times are portrayed in a plot of more or less complexity.'

This may sound a bit like stating the obvious, but there are some important points contained in this very concise definition.

The novel is fictitious – *fiction*, as we often refer to it. 'Fiction' is a very familiar concept, one with which most people are familiar and which few find problematic. 'Telling stories' is a function that is so central to our culture that we pay it little analytical attention: we are accustomed to the need to distinguish between true reports and fictional stories – 'make-believe' or 'pretend' – from a very early age, and in this case 'we' does seem to refer to all human beings. Fiction is much wider than the novel or, indeed, than prose fiction: jokes, imitations and parodies, songs, and narrative poems can all be described as fictions – and wider, non-literary usages include such things as legal fictions and (perhaps) folk tales and urban myths. Now although all of these share some common elements, we should beware of assuming that they are all the same: as I will argue later on, even fictional stories may vary so significantly that it makes sense to separate out one particular set of fictional stories and distinguish them by means of the term 'novel'. Even so, I would want to argue that the specific tradition or set of traditions which we refer to as the novel is made possible by a far more widespread and fundamental reliance upon fiction in human society; coming to terms with the novel happens in easy stages in literate societies and appears to build upon universal habits of fantasy, play, rôle-playing and joking.

Like many very familiar cultural forms, fiction turns out to be tantalizingly hard to define once we make the attempt so to do. We can start with a number of seeming paradoxes: fictions are not true, but they are not lies; they typically describe that which is not real but which is nonetheless not totally unreal (and they can include references to real people and events without jeopardizing their fictional status); they are designed to get readers or listeners to respond 'as if', but not (normally) to deceive him or her; it is known that they (sometimes) describe people who do not exist or events which have not happened, but they do produce real emotions, important reflections, and even (although this is the subject of much disagreement) altered behaviour in the real world. Moreover, it seems at least possible that fiction is not a wholly stable category and that the fictional elements in, say, the myths of a non-literate culture, operate differently from the fictional element in Charles Dickens's *Dombey and Son* (1847–8) or Salman Rushdie's *The Satanic Verses* (1988). And this last example reminds us that the connections between fiction and the world may not only be complex, but may vary from reader to reader or culture to culture. Finally, we should remember that a fiction does not have to take the form of a story or a narrative.

In their book *Truth, Fiction, and Literature: A Philosophical Perspective*, Peter Lamarque and Stein Haugom Olsen provide us with a number of useful distinctions with which to approach this complex topic. First, they differentiate between what they call the 'object' and the 'description' senses of the term 'fiction', arguing that a fictional character is a fiction in the object sense while a work of fiction is a fiction in the description sense.

> To say of a thing that it is fictional is to suggest that it does not exist, the implied association being between what is fictional and what is *unreal*. To say of a description that it is fictional is to suggest that it is not true, the implied association being between what is fictional and what is *false*.[1]

Second, they distinguish between fictional content and fictional mode, illustrating the difference by reference to Konrad Kujau's forgery which was presented to the world as Hitler's (genuine) diaries. As they remark, while it would be reasonable to refer to the content of this forgery as fictional, 'we should surely hesitate to call the finished product a "work of fiction" given the mode of its presentation'.[2] These distinctions allow them to focus not on the structural or semantic properties of sentences 'but on the conditions under which they are uttered, the attitudes they

[1] Peter Lamarque and Stein Haugom Olsen, *Truth, Fiction, and Literature: A Philosophical Perspective* (Oxford, Clarendon Press, 1994), p. 16.

[2] Peter Lamarque and Stein Haugom Olsen, *Truth, Fiction, and Literature*, p. 17.

evoke, and the role that they play in social interactions'.[3] Such an approach leads to a stress on fictional storytelling as a convention with institutional resources or, to put it another way, as a more or less universally understood custom or family of customs within a culture.

My own feeling is that a comparison with the concept of play can be particularly instructive here. Anyone who has watched a child grow and mature will know that he or she attempts to understand and master the world and his or her relation to it through modelled performances which we call play. However much we may like to think of play as sharply distinguished from the real, workaday world, relegated to 'free time' or 'relaxation', there seems little doubt that the rules of games can function as a model of the social and material restraints which we meet with in the world, and that play can help us to internalize these aspects of the world and of our operating within it.

Fictional stories, too, present us with models of (especially) the social world which we can empathize with and observe because we are not constrained to act. According to Lamarque and Haugom Olsen, readers 'often fantasize with fictive content, "filling in" as the whim takes them, and no doubt some genres of fantasy actively encourage this kind of whimsical response'. This is rather stern, and it becomes more so as the two authors move on to insist upon the need to distinguish between authorized and non-authorized responses – with 'the content itself and its presentation' as the source of this authority.[4]

I call this a rather stern pronouncement in spite of the fact that as a university teacher I spend a lot of time telling my students that they need to be able to justify their responses and interpretations by reference to textual and other evidence, and this is because, although academic discussion of literature has to insist upon such standards in order to establish ground-rules for discussion and debate, I do not think that readers (as against students or critics) of novels are – or ever have been – bound to remain within the boundaries of the authorized. Indeed, were it possible to render 'whimsical responses' impossible, then I suspect that novel-reading would be a lot less popular than it is today. I agree that certain aspects of literary response are like those forms of play which are bound by very rigid rules: football or chess, for example. But reading novels also resembles the child's make-believe games, those games where it is the imagined rôles and situations which offer the only constraints. To

---

[3] Peter Lamarque and Stein Haugom Olsen, *Truth, Fiction, and Literature*, p. 32.

[4] Peter Lamarque and Stein Haugom Olsen, *Truth, Fiction, and Literature*, p. 89.

a certain extent Peter Lamarque himself clearly accepts this point, as in a later book he argues that fictive storytelling 'disengages standard conditions of assertion, it invites imaginative rather than belief-based involvement, it creates worlds and characters, and it encourages participation, not a concern for correspondence with the facts'.[5]

This is not to argue that works of fiction impose no constraints on what we imagine while we are reading them. Indeed, it is precisely because our responses, our imaginings, have to confront and negotiate with what cannot be changed in a novel that the reading of fiction is able to model what it is like to live in the social and material world. We cannot read on through Emily Brontë's *Wuthering Heights* (1847) pretending that Catherine does not really die but is only hiding somewhere, so that she can re-emerge outside Lockwood's window and give him a shock. But one of the things which games do is to allow us to act out *alternative* moves, modes of behaviour, or whatever, in a way impossible in the real world where we often have to act quickly and irrevocably – and this, I believe, is also true of novels and short stories. Thus although I would reach for my red pen if I received an essay from a student which argued that Catherine never really died but lived on the moor as a hermit, it seems to me that wondering what would have happened had Catherine not died is an absolutely normal part of reading a novel such as *Wuthering Heights*. Now it is certainly true that what it may be interesting to wonder during one's reading of a novel is very often not worth discussing once one has finished the novel. But this should not lead us to reject such wondering – or comparable indulgences of fantasy, imagination, daydreaming – as illegitimate, 'wrong', unauthorized, or whatever. It is from the soup of such responsive processes that our interpretative and analytical understanding is distilled.

The play of children is paradoxically double-sided: on the one hand it clearly forms part of that process we call socialization, through and as a result of which individuals acquire and internalize rules of socially sanctioned behaviour. But at the same time it allows participants to test out limits, to question rules, to weigh up different ways of behaving. This is why I tend to react with suspicion to too great a stress on 'authorized responses', even though this is a term I do find myself using when marking student essays. In the world we inhabit while we read a novel there is one factor for which the author can never fully allow – and that

[5] Peter Lamarque, *Fictional Points of View* (Ithaca, Cornell University Press, 1996), p. 144.

is the uniqueness of each reader and each reading. And as a result there can be no single, final, true, authorized response to a complex novel. Response is interactional, and in each reading one of the interacting elements is different.

Now this book is called *Studying the Novel*, not *Responding to the Novel*, and my emphasis throughout it is on those institutionally approved ways of categorizing, analysing, and interpreting novels and short stories. I am concerned, in other words, with the ways in which those private, individual and utterly varied reading responses people have can be introduced into society and allowed to converse to others. If our talk to others about novels is to be more than a semi-comprehensible outpouring of our highly idiosyncratic reading experiences then we have to anchor it to things that we can all share: textual details, scholarly procedures, and so on. But we should be very careful to avoid the assumption that such public and 'objective' information is all that matters.

Compare the difference between dreaming and reporting upon one's dream to another person. The report has to be such that another person can follow it, and if we believe that dreams are worth discussing, analysing, and interpreting then these processes can only take place on the basis of understandable reports. But it would be a fundamental error to assume that the report 'is' the dream: our report may state that we were terrified, but in the dream we *are* terrified; we can laugh at the report of the terror, but the experience of the terror is not funny.

In the same way, I believe, reading fiction produces responses and mental processes which are too complex ever to be fully captured in discussion of novels and short stories. This does not make such discussion pointless: learning to study the novel helps us to refine and extend our reading experiences, and it allows us to turn the experience of reading into knowledge and self-knowledge through ordered and rational discussion. But we should never be ashamed to admit that the reading of stories is a mysterious process, only a part of which can be held on to and brought into the light of day through analysis and interpretative discussion. We actually know relatively little about what happens when people read novels: most academic discussion is of texts and works rather than of reading experiences. We cannot avoid this, but we can and should avoid the pretence that there is nothing else. We need to retain that sense of the mystery of reading, even while we attempt to plumb its hidden depths and explore its puzzling complexities.

We will have to come back to some of these problematic issues later, but for the time being we can agree that the novel depicts imaginary characters and situations. A novel may include reference to real places, people and events, but it cannot contain only such references and remain a novel. However, even though its characters and actions are imaginary they are in some sense 'representative of real life' as the dictionary definition has it; although fictional they bear an important resemblance to the real. What exactly this resemblance is has been a matter of much discussion and dispute amongst literary critics, and it is arguable that it varies in kind from novel to novel. But this resemblance to *real* life is one of the features that distinguishes the novel from other forms such as the epic and the romance, however much we recognize that the term 'real life' is a problematic concept. Later on in Chapter 5 we will see that the term 'realism' is one which, although arguably indispensable to discussion of the novel, requires careful definition and use.

The Russian critic M.M. Bakhtin has reminded us of the important if paradoxical fact that the novel's association with 'real life' is far from excluding fantasy and imagination from the experience of reading fiction. Because there is not normally any fundamental or uncrossable distance between our everyday life and the life represented in the novel, when we read the novel we are able to enter its world in a way that the reader of (or listener to) the epic could not do. It is common knowledge that people can 'lose themselves' in a novel. In other words, even though the novel presents us with a recognizable world, we exercise our fantasy and our imagination to live within this world for short periods of time. According to Bakhtin, this form of escape from the everyday into what we can call a *represented everyday* can be taken to an extreme: 'we might substitute for our own life an obsessive reading of novels, or dreams based on novelistic models'.[6] Most present-day readers are probably still familiar with that objection to the novel that associates our ability to lose ourselves in a story with the assertion that novels are 'just entertainment', or 'a means of escape'. The flaw in this argument is that it does not take into account the possibility that, even if readers *do* lose themselves completely in reading a novel (and the reality is probably more complex than this), this does not necessarily mean that we do not mull over and learn from our reading experiences once we return to our everyday reality after finishing a novel.

[6] M.M. Bakhtin, 'Epic and Novel', (trans.) Caryl Emerson and Michael Holquist. In M.M. Bakhtin, *The Dialogic Imagination: Four Essays*. Michael Holquist (ed.), (Austin, University of Texas Press, 1981), p. 32.

Bakhtin also points out that not only can the reader of the novel enter its world, but so too can its author.[7] This is one reason why it can be complex and problematic to determine to what extent opinions expressed in a novel do or do not represent the views of its real-life author.

The world of the novel is so familiar to us that we can on occasions recognize its distinctive qualities only by contrasting it to other genres. Bakhtin points out, for example, that '[t]he world of the epic is the national heroic past', it is based upon a 'national tradition (not personal experience and the free thought that grows out of it)', and 'an absolute epic distance separates the epic world from contemporary reality'. The novel, 'by contrast, is determined by experience, knowledge and practice (the future)'.[8] To simplify: we live *with* the characters of a novel, whereas we look *at* the happenings that take place in an epic account. The novel presents us with stories, experiences that are new: whereas those who experienced the epic knew what was to occur, when we pick up a novel we hope that our curiosity will be aroused by our wanting to know 'what happens'. As I will argue in Chapter 2, this difference has a lot to do with the different sorts of society from which these two genres emerge.

The novel is in *prose* rather than verse, although the language of novels may often strike us as very 'poetic' on occasions. And although it would be a serious mistake to assume that the language of a novel was identical to that of ordinary speech or of most non-literary writing, nevertheless the fact that the novel is in prose helps to establish that sense of 'real life' – of recognizable everyday existence – that is the preserve of the genre. Generalization is dangerous, but prose has the potentiality of being a more transparent medium than, for example, verse: reading a novel our attention is not naturally drawn to considering the language as language. The tendency is more for us to 'look through' the language of the novel at what it describes and evokes than to 'look at' it – although novelists can write prose that arrests our attention as language, and many novelists are known for the manner in which they draw the reader's attention to their use of language. It is certainly the case that in the course of studying novels we must learn to pay more overt attention to their language than does the average casual reader.

According to our definition, a novel is concerned with 'the real life of past or present times'. There are of course many works which we would not hesitate to call novels which are set in the future – and this was

[7] M.M. Bakhtin, 'Epic and Novel', p. 27.

[8] M.M. Bakhtin, 'Epic and Novel', pp. 13, 15.

also the case at the time that the OED definition was written. It is probable that it is the novel's association with what seems 'real' to us, with a recognizable human life in existing or past cultures, that has prompted the exclusion of 'the unreal' (however defined) from the dictionary definition. Even novels which have a significant dependence upon the fantastic typically seem still to be much more connected to the real life of past or present times than do works belonging to genres such as epic or romance. In Chapter 2 I will outline some disagreements about how we define the novel which, although long-standing, have come to a head in recent years. At the centre of such disagreements is precisely the issue of whether an insistence upon 'real life' is a helpful component in any attempt to come up with the sort of neat definition that the OED presents us with. Those who believe that it is not argue that many modern novels are not primarily concerned with 'real life', and that the insistence upon limiting our sense of the novel in this way cuts the modern novel off from what is often referred to as 'the novel of the ancient world'. Although such arguments are important and should be considered by anyone studying the novel, in the end I have to admit that I am not convinced by them. But we can return to this issue later on.

The novel is a *narrative*: in other words it is in some sense a 'telling' rather than an 'enacting', and this distinguishes it in an important sense from the drama. Of course novels can contain very dramatic scenes, and often the reader may forget that what we learn of character and event is not experienced directly through a presentation or an enactment (as in the theatre or the cinema) but mediated through a particular telling, a narrative source. Take the opening of Henry James's short story 'The Lesson of the Master' (1888):

> He had been told the ladies were at church, but this was corrected by what he saw from the top of the steps – they descended from a great height in two arms, with a circular sweep of the most charming effect – at the threshold of the door which, from the long bright gallery, over-looked the immense lawn.

If our first impression is that we are witnessing a scene directly, a second glance will confirm that we are actually being *told* about the scene. The telling is such that we can visualize what is described – that is often the mark of an accomplished narrative – but we see what is first selected and then pointed out to us by a narrator or a narrative voice or source. Of course, this selection and presentation is fictional – the author is actually *creating* what we see. But as we read we are given the sense that the scene's potentially infinite complexity is reduced, ordered and explained to us through the organizing and filtering consciousness of a narrator.

In a filmed version of this scene the different members of the cinema audience could choose to concentrate upon different aspects of what unfolded before them; the reader of James's short story, in contrast, 'sees' what James's narrator chooses that he or she shall see. Imagine that we were presented with a filmed version of the above scene. Would we necessarily think of the ladies as forming 'two arms'? Would we find the effect of the 'circular sweep' 'most charming'? It is often assumed that a film or dramatization of a novel can include more than the original, but the loss of a directing and organizing narrative source very often means that although such filmed or dramatized versions can be visually rich and dramatically immediate, they lack much of the verbal and artistic richness of their originals. The novelist B.S. Johnson has commented with regard to film that it 'is an excellent medium for showing things, but it is very poor at taking an audience inside characters' minds, at telling it what people are thinking'.[9]

Drama typically 'tells' less and 'shows' more than narrative, and although theorists of fiction (including, importantly, Henry James) have given such showing a higher standing than telling, the above quotation reveals how much even James himself relied upon telling. How, for example, would the director of a filmed version of the scene in question convey that 'he' had been told something earlier, while leaving us absolutely in the dark at this point with regard to who 'he' is, what he looks like, and who told him about the ladies being at church?

There is an important theoretical point to be made here. The teller of a tale can often choose to inform the reader of just what he or she wants – no more and no less. But dramatic displays cannot be fine-tuned in quite this way, and are, inevitably, rich in peripheral information. James may have wanted us to see his ladies in a collective, more 'aesthetic' manner at this stage of his story. But it would be hard for a filmed version of the scene to prevent the viewer from seeing them also as individuals. James chose to restrict the information given the reader about the setting in which this scene takes place: for a filming of the scene to do the same would require immense technical skill on the part of the director.

---

[9] B.S. Johnson, Introduction to *Aren't You Rather Young to be Writing Your Memoirs?* (First publ. 1973, repr. in Malcolm Bradbury [ed.], *The Novel Today: Contemporary Writers on Modern Fiction* [revised edition, London, Fontana, 1990]), p. 166.

One word of caution is needed at this point. I have said that the novel is a narrative, a telling. But the term 'narrative' can be used both with a narrow and specific meaning and also with a rather wider one. The more narrow meaning restricts narrative to the telling of events, the recounting of things that happen. (Clearly events can just as well be mental as physical: a sudden dawning of insight is as much an event as is the outbreak of the First World War.) Those with an interest in etymology may like to have their attention drawn to the fact that the words 'tell' and 'recount' are both related to the action of counting, or enumerating (think of a bank teller, who counts money). And as 'tell' is etymologically related to 'tale' we can say that behind the act of narrative itself lies the idea of communicating things that can be counted. (Compare the French *conte*, which I discuss briefly below.) If we limit ourselves to such a narrow definition of narrative, therefore, we must remember that not everything in a novel is narrative. *Description* (of a beautiful scene) or *argument* (about, say, the existence of God) are neither of them examples of narrative in this strict sense; they do not involve the telling or enumerating of discrete events or countable entities. But the term is also used in a rather wider sense to include everything that comes within the purview of a particular telling or recounting. From this perspective everything within the pages of a novel is narrative, including description and argument. If you think about it, much the same is true of our use of the word 'fiction'. After all, many statements in novels are literally true, yet we do not restrict use of the word 'fiction' to those parts of novels of which this cannot be claimed.

The novel has *characters*, *action(s)*, and a *plot*: it presents the reader with people who do things in a total context ruled over by some sort of connective logic: chronology, cause-and-effect, or whatever. There is, moreover, in most novels a connection between these three elements such that they form some sort of unity. A poem does not have to contain characters or a plot – or, indeed, any action – but it is only very rare novels which dispense with one of these elements, and in such unusual cases it is often a matter of dispute as to whether the net result is recognizable as a novel. Some recent narrative theorists have preferred to talk of *actors* rather than characters, pointing out that a work of science fiction might be based on rôles filled by non-human participants. This would be an extreme case, but one can further point out that novels very often personify non-human forces and make them responsible for important actions, so that the association of the term 'character' with 'human individual' may be seen as somewhat limiting.

We have of course to give the term 'action' a relatively broad meaning in the present context. Much recent fiction involves significant concentration upon what has been termed 'inner action': events taking place in a character's consciousness rather than in the social and physical worlds outside. It is nevertheless revealing that works consisting only, or mainly, of such inner action are often termed 'prose poems'. It is as if our view of the novel requires that it contain a significant amount of action more traditionally defined.

In Chapter 6 I will be talking about the important distinction between 'story' and 'plot'; at this stage I will limit myself to stating that the concept of plot involves some sense that the actions we have talked about represent a whole rather than merely a succession of unconnected events. Here again care is needed, for as my comments on the picaresque novel in Chapter 3 make clear, such novels (and others) can be characterized by a very *episodic* structure. Jerome J. McGann's comment that '[Dickens's] early work is far more episodic than the later, so much so that many would be reluctant to call *Pickwick Papers* [1836–7] a novel at all'[10] thus has a representative force: nowadays readers tend to like their novels to have plots that unfold rather than episodes that merely succeed one another in an order that is without significance.

Ian Watt has argued that the novel is distinguished by the fact that unlike the works of the great English and Classical poets and dramatists its plots are generally not taken from traditional sources.[11] If we compare the modern novel with the prose fictions of antiquity, or the medieval romance, Watt's point is certainly justifiable. On the other hand even the modern novel has recourse to certain recurrent plots and story-lines (see the discussion of plot on p. 125.) As a result, many modern novels manifest an important and productive *tension* between their fidelity to the random and unpredictable nature of everyday life, and their patterning according to the demands of certain predetermined but widely different structures, from 'tragedy' to 'repetition' or 'the journey'.

Finally, the novel is of a certain length. A poem can be anything from a couplet to a thousand pages or more, but we feel unhappy about granting the term 'novel' to a tale of some forty or fifty pages. Of course it is not just a question of length: we feel that a novel should involve an

[10] Jerome J. McGann, *The Textual Condition*, (Princeton, Princeton University Press, 1991), p. 82

[11] Ian Watt, *The Rise of the Novel* (Harmondsworth, Penguin Books, repr. 1963), p. 14.

investigation of an issue of human significance in such a manner as allows for complexity of treatment, and by common consent a certain length is necessary to allow for such complexity. In practice, therefore, we usually refer to a prose narrative of some twenty or thirty pages or less as a short story, while a work that seems to hover on the awkward boundary between 'short story' and 'novel', having a length of between forty or fifty and a hundred pages, is conventionally described as a novella (plural: novelle or novellas).

These terms have achieved the relative fixity of meaning they now enjoy only comparatively recently. At the beginning of the twentieth century Joseph Conrad reacted against a critic who referred to his having written a short story, and clearly himself preferred the terms 'tale' and 'story' to describe what modern critics refer to as his short stories. (A tale is often distinguished from a short story by reference to the greater interest in 'what happens' that it encourages in the reader, which is presumably one of the reasons, length apart, why Conrad's own *Heart of Darkness* [first book publication 1902], which consists mainly of what its frame narrator tells us is 'one of Marlow's inconclusive experiences', is *not* described as a tale.) The situation is complicated by the fact that *Lord Jim* (1900), which no-one today would deny was a novel, is also described as a tale in its subtitle in standard editions, although Conrad started off by referring to a short early version as 'A Sketch', and an edition published in his lifetime described the work as 'A Romance'. Such changes suggest an attempt to escape the limitations of a set of generic requirements or expectations.

Two other terms that may be encountered are the French loan-words *conte* and *nouvelle*. The *conte* was originally distinguished from the novel by its (often quirky) wit and its more fantastic, less realistic nature, and by a strongly didactic element normally focused in a moral point or use of allegory. Typical examples are two works both first published in 1759: Samuel Johnson's *Rasselas* (first published as *The Prince of Abissinia. A Tale*) and Voltaire's *Candide*. For the past century or so, however, the term in English has become roughly synonymous with 'short story', although Frank O'Connor (whose views on the short story I will discuss further in Chapter 4) categorizes Robert Browning's poem 'My Last Duchess' (1842) and Ernest Hemingway's 'Hills Like White Elephants' (1927) as *contes*.[12]

[12] Frank O'Connor, *The Lonely Voice: A Study of the Short Story* (London, Macmillan, 1963), p. 26.

A *nouvelle* is characterized by a concern with a single episode or state of affairs, although its treatment of this (which conventionally moves to a surprise ending of one sort or another) may cover many pages. The genre is generally agreed to achieve a recognized form in the middle of the sixteenth century, with the publication of *Nouvelles Récréations et Joyeux Devis*, a collection ascribed to Des Périers. The term can be a little treacherous as in some usages it has come to be seen as synonymous with *conte*, while in others it is used interchangeably with *novella*.

In my next chapter I will discuss the case made by a number of recent literary historians that terms such as 'novel' and 'romance' achieved a fixity of meaning only in the nineteenth century. We need to remember that the same may be true of terms such as 'novella' and 'short story' in the twentieth century. Furthermore, what we understand by the term 'novel' has necessarily been enlarged by the novels that have been written in our time, and it will no doubt go on to be further enlarged. This is doubtless true of most of the generic terms used in literary criticism. But, because the novel's rejection of traditional forms and conventions as well as its ability to respond to and incorporate aspects of the changing world around it into itself are fundamental to its existence, the novel is *generically* characterized by a sort of permanent revolution of both its form and its content.

2 History, Genre, Culture

In the first two editions of this book I wrote as if there was little disagreement about what constituted a novel. I did, it is true, refer to the 'running debate' between those who see prose fiction as 'a universal and ancient form with a continuous history',[1] and those who prefer to emphasize the distinctiveness of the prose fiction which emerged in the early eighteenth century, and who speak of the novel as a new form which had its birth then. Even so, my account implied that although 'the novel' as category might be fuzzy round the edges, and might be in a state of permanent development and self-reconstruction, nevertheless this new literary genre emerged in the Europe of the late seventeenth and early eighteenth century and was clearly different from earlier forms such as that of the romance.

This, it has to be said, remains the consensus view. But it is not without its opponents, and recently the 'running debate' has heated up somewhat, especially following the publication of a weighty study *The True Story of the Novel* (1996), by Margaret Anne Doody. Doody reminds her readers that the distinction between novel and romance is one that is not reflected in all languages in the way that it is in English, and she proposes that it 'has outworn its usefulness, and that at its most useful it created limitations and encouraged blind spots'.[2] For her, the novel 'as a form of literature in the West has a continuous history of about two thousand years', a fact which recent scholarship has denied or obscured.[3] As one might expect, in order to press her case she advances a rather

---

[1] Maximilian E. Novak, review of J. Paul Hunter, *Before Novels: The Cultural Contexts of Eighteenth-century English Fiction. Times Literary Supplement*, no. 4582, January 25, 1991, p. 8.

[2] Margaret Anne Doody, *The True Story of the Novel* (New Brunswick, New Jersey: Rutgers University Press, 1996), p. xvii.

[3] Margaret Anne Doody, *The True Story of the Novel*, p.1.

wider and more all-embracing definition of the novel from that which I quoted from the OED at the start of my first chapter; for her, a work is a novel if 'it is fictional, if it is in prose, and if it is of a certain length'.[4]

There is no doubt that if one defines the novel in this way then Doody is quite correct to see it as a literary form with a continuous history of two thousand years, and it is certainly the case that her argument has the virtue of calling our attention to continuities in that tradition of fictional prose which now exists primarily in the form of the modern novel. Now if we want to adopt Doody's definition we may do so: we should certainly try to avoid that error known as 'nominalism' which assumes that we can change reality by means of the names which we give to sections of it. For myself, I still believe that what happened to prose fiction in the course of the eighteenth century in Europe represents such a radical change that what to avoid debate we can call 'the modern novel' merits and repays separate study. As Michael McKeon points out, it was the first modern 'novelists' who themselves affirmed that 'a distinct new form' had emerged: Richardson believed he had introduced a new 'species' of writing while Fielding claimed a new 'kind' or 'province'.[5]

Doody's very well documented argument has the effect of drawing our attention to continuities – which is what she intends – but its descriptions of 'ancient novels' has, too, another effect which I feel sure is not intentional: it allows us to confront the fact that these ancient prose fictions are radically different from what we know as the novel. Readers are invited to study Doody's arguments in her book, to read some of the ancient fictions to which she refers, and to decide for themselves how they wish to use the term 'the novel'.

But how the modern novel did emerge is not a simple process to register. In some ways it is a little like reading one of those historical accounts of the emergence of the human species; all sorts of near-misses occur, dead-ends of development peter out, and then – miraculously – all the required ingredients come together and human beings emerge. And just as researchers are still arguing about 'disputed ancestors' of the human species (are we or are we not descended from this particular ape-like creature, and do these fossil remains stem from a human being or something else?), so too literary critics and historians are by no means in total agreement about who the ancestors of the novel are, or, indeed, quite

---

[4] Margaret Anne Doody, *The True Story of the Novel*, p. 16.

[5] Michael McKeon, *The Origins of the English Novel 1600–1740* (Baltimore, Johns Hopkins University Press, 1987), p. 410.

when the novel came into existence. (The argument about when apes end and humans begin is paralleled by arguments about whether various forms of early narrative should or should not be described as 'novels'.)

The 'running battle' I mention above involves, among other things, detailed debates about the extent to which those 'novel' works produced by writers such as Defoe, Richardson and Fielding in the first half of the eighteenth century were in direct line of descent from the chapbooks and the French fiction of the late seventeenth century, hundreds of titles of which were translated into English and published in England at this time.[6] It should not, however, prevent us from recognizing that, even if the novel can be said to emerge as a new literary genre in the eighteenth century, it still owes much to traditions and works, literary and non-literary, from earlier times. No serious student of the novel would deny, for example, that its development in the eighteenth century was pro-foundly influenced by works such as François Rabelais's *Pantagruel* and *Gargantua* (1532 and 1534) and Miguel de Cervantes's *Don Quixote* (1605–15) – whether or not one agrees to describe these works as novels themselves.

As with the problems attendant upon tracing the emergence of *Homo sapiens*, claims concerning parentage are often controversial. One of the things which sets the novel apart from many other literary genres is its ability to incorporate the most disparate elements from human life and experience in itself. It would be a serious mistake to assume that to trace the novel's descent we need to examine only a sequence of written forms, or of oral narratives. We should rather picture a family tree in which certain lines of descent involve various written forms, certain involve a succession of oral narrative forms, but others involve a wide range of very different elements: introspective self-analyses, both in diary and in unspoken form; joke-telling; sermons; travel accounts; letters. What distinguishes the novel is, among other things, the heterogeneity of its ancestry, a heterogeneity that Bakhtin has argued is mirrored in the variety of different 'voices' to be found in any single novel. He claims (and this is by no means uncontroversial), that, however many contradic-tions and conflicts are developed in a poem, the world of poetry is always illuminated by 'one unitary and indisputable discourse'. The novel, in

---

[6] Maximilian E. Novak, review of J. Paul Hunter, *Before Novels: The Cultural Contexts of Eighteenth-century English Fiction*, p. 8.

contrast, is he argues characterized by that 'variety of individualized voices' that is 'the prerequisite for authentic novelistic prose'.[7]

One of the impulses that often moves those who compile family trees is a desire to prove that all the discovered ancestors were respectable and legitimate. In the case of the novel, we have to be prepared to accept the possibility that many of its ancestors were neither. As J. Paul Hunter has put it:

> I join the debate here by arguing that the emerging novel must be placed in a broader context of cultural history, insisting that popular thought and materials of everyday print – journalism, didactic materials with all kinds of religious and ideological directions, and private papers and histories – need to be seen as contributors to the social and intellectual world in which the novel emerged.[8]

There is, at any rate, general agreement that *fiction* is central to the novel and its development. Martin Seymour-Smith has suggested that the first fiction was mythology and folklore.[9] One could go further, and claim that, because language allows us to imagine what it would be like if that which is the case were not the case, fiction in the sense of imagination appears when human beings start to use a language that is comparable to our own. From this perspective, even the use of metaphor (which is fundamental to all language) can be seen as a sort of fictional device: by saying, 'He is a lion', we are actually saying, 'Let us imagine that he is a lion'. Mythology and folklore extend this use of metaphor to coherent narrative accounts. *Fictional narratives* can be found almost as far back as we have written records, but these lack many of the characteristics that today we associate with the modern novel. First, they are often in verse rather than prose. Second, they do not concern themselves with 'the real life of past or present times' but portray the experiences of those whose lives can only be said to resemble 'real life', however it is defined, in

---

[7] M.M. Bakhtin, 'Discourse in the Novel'. In M.M. Bakhtin, *The Dialogic Imagination: Four Essays* (Austin, University of Texas Press, 1981), pp. 286, 264. 'Voice', for Bakhtin, includes ideological and other elements: it is not just a question of the physical voice.

[8] J. Paul Hunter, *Before Novels: The Cultural Contexts of Eighteenth-century English Fiction* (London, W.W. Norton, 1990), p. 5.

[9] Martin Seymour-Smith, 'Origins and Development of the Novel'. In Martin Seymour-Smith (ed.), *Novels and Novelists: A Guide to the World of Fiction* (New York, St. Martin's Press, 1980), p. 11.

extremely convention-governed and indirect ways (gods or mythical heroes, for example). Third, their characters tend not to be individualized – to the extent that even the use of our familiar term 'character' is problematic. (We do not feel too comfortable calling a purely stereotyped or conventionally portrayed individual a character precisely because today the term 'character' is associated with distinctive traits, with some particularity or individuality.) The etymology of 'character' is very revealing: the word comes from a Greek word for an instrument used for marking or engraving, just as 'style' is etymologically related to the word stylus and is also associated with a cutting instrument. As a result, the word's literal meanings – as the OED will confirm – are related to *distinctive* and significant marks (think of the obsolescent use of the word to mean handwriting). From this we get the familiar OED definition: 'A personality invested with distinctive attributed and qualities, by a novelist or dramatist.' Our sense of human individuality, this etymology suggests, is closely related to making marks – to *writing*. Writing allows human beings to develop individualities, to be and to know themselves to be, different, particular, unique.

Now it seems probable that only a very small proportion of the narratives which existed in the ancient world has survived: even the scattered and fragile fossils of our ancestors are more durable than the remains of early written records, and only a tiny proportion of all narratives have ever been written down. (Only a tiny proportion are still written down today.)

Another problem is that, in comparison to other literary genres, the conventions governing the novel appear to be extremely flexible. Thus some novels have relatively unindividualized characters; others seem a long way away from the real life of past or present times; and some (even today) seem much closer to poetry than to the prose of, for example, a newspaper report.

There is no better example of the problems attendant upon generic classification than that of the *romance*, universally agreed to represent one of the most important traditions contributing to the emergence of the novel. (The term 'roman', as Margaret Anne Doody reminds us [see p. 16] is the equivalent of 'novel' in many modern European languages, while the word 'novel' comes, confusingly, from the Italian 'novella' – 'small new thing'.)

Nowadays we have no difficulty in distinguishing the novel from the romance, although the existence of modern 'romances' reminds us that the novel today also includes many highly formulaic and stereotyped sub-genres. But the traditional chivalric romance which developed in twelfth-century France depicted not epic heroes but a highly stylized and idealized courtly life founded upon rigid but sophisticated conventions of behaviour. Like the epic (which it displaced) it often involved supernatural elements – another factor which in general terms distinguishes it from the modern novel. The distinction with which we are familiar can be found in the eighteenth century, the century in which the novel appears in its recognizable modern form in England. Geoffrey Day quotes the following from a review of Fanny Burney's novel *Camilla*, which appeared in *The British Critic* for November 1796:

> To the old romance, which exhibited exalted personages, and displayed their sentiments in improbable or impossible situations, has succeeded the more reasonable, modern novel; which delineates characters drawn from actual observation, and, when ably executed, presents an accurate and captivating view of real life:[10]

As Day notes, 'throughout the eighteenth century "romance" was seen by many as a term suggesting excessive flights of fancy',[11] and he quotes a number of examples of such a usage. But he goes on to point out that this neat definition, one which served to distinguish the romance from the novel, was not universally accepted: many writers tended to lump the two terms together relatively uncritically. He comments:

> [T]hough there is a tradition of clarity of definition, there is vastly more evidence to show that those works now commonly referred to as 'eighteenth century novels' were not perceived as such by the readers or indeed by the major writers of the period, and that, far from being ready to accept the various works as 'novels', they do not appear to have arrived at a consensus that works such as *Robinson Crusoe*, *Pamela*, *Joseph Andrews*, *Clarissa*, *Tom Jones*, *Peregrine Pickle* and *Tristram Shandy* were even all of the same species.[12]

(And in case we should be tempted to feel superior to these dull individuals who were incapable of recognizing the birth of a new genre,

---

[10] Geoffrey Day, *From Fiction to the Novel* (London, Routledge, 1987), p. 6. Day's study is a mine of useful references which display how eighteenth-century British writers saw what we now recognize as the developing novel.

[11] Geoffrey Day, *From Fiction to the Novel*, p. 6.

[12] Geoffrey Day, *From Fiction to the Novel*, p. 7.

it is salutary to stop and think about the enormous formal variety that this list of works represents.)

The eighteenth-century man of politics Lord Chesterfield, writing to his son about the novel, describes it in terms that suggest that it is to be distinguished from the romance on the grounds of its reduced length:

> I am in doubt whether you know what a Novel is: it is a little gallant history, which must contain a great deal of love, and not exceed one or two small volumes. The subject must be a love affair; the lovers are to meet with many difficulties and obstacles, to oppose the accomplishment of their wishes, but at last overcome them all; and the conclusion or catastrophe must leave them happy. A Novel is a kind of abbreviation of a Romance; for a Romance generally consists of twelve volumes, all filled with insipid love nonsense, and most incredible adventures.

It is true that Chesterfield goes on to suggest a distinction that is closer to that by which we would, today, distinguish novel from romance:

> In short, the reading of Romances is a most frivolous occupation, and time merely thrown away. The old Romances, written two or three hundred years ago, such as Amadis of Gaul, Orlando the Furious, and others, were stuffed with enchantments, magicians, giants, and such sort of impossibilities, whereas the more modern Romances keep within the bounds of possibility, but not of probability.[13]

Chesterfield's comments are interesting not just because of the manner whereby he distinguishes the romance from the novel, but also because of the light that they throw upon the relatively low status that the novel enjoyed in its early years. We should remember that even when the study of English literature was introduced into British universities in the early part of the twentieth century it was expected that this would in the main be limited to drama and poetry. Chesterfield was of course an aristocrat, with an education that had steeped him in classical learning. As I will be pointing out, in the century of its birth the modern novel was associated not with the aristocracy but with the rising middle class, whose members often lacked Chesterfield's classical learning.

---

[13] Charles Stokes Carey (ed.), *Letters Written by Lord Chesterfield to his Son* (London, Wm Reeves, 1912), volume 1, p. 90. (Original letter in French). The letter is undated but is probably written around 1740–41.

As Day demonstrates, Chesterfield was not alone in insisting upon length as a distinguishing feature that set the (shorter) novel apart from the (longer) romance. He notes that in his *Dictionary* (1755), Samuel Johnson defined a novel as 'A small tale, generally of love', adding that the definition was not without its contemporary critics.

Thus although for our immediate purposes it is important to distinguish the novel from the romance, the novel superseded the romance by, to a certain extent, incorporating some important elements of the romance in itself, and the dividing-line between romance and novel was a good deal less clear to many early novelists than it seems to us. (As Day points out, we should remember that Daniel Defoe, for example, never referred to his own works as novels. As some of these were presented to the public in a form that suggested that they were true accounts, this is perhaps not surprising.) There is, moreover, a recognizable line of development in the novel which often reminds us of aspects of the romance, one which proceeds from the gothic novel through science fiction to fantasy works such as *The Lord of the Rings* (1954). (See Chapter 3.) Michael McKeon points out that 'even though Defoe, Richardson, and Fielding explicitly subvert the idea and ethos of romance, they nonetheless draw upon many of its stock situations and conventions', and he suggests that the 'general problem of romance in all three novelists is related to the particular problem of spirituality, equally antithetical to the secularizing premises of formal realism, in Defoe'. In addition, McKeon draws attention to the fact that the romance exhibits certain characteristics of formal realism, and that it continues to co-exist with the novel during the period of the latter's emergence.[14]

In spite of all these reservations and disclaimers it is the case that what we now know as the novel does stand apart from what we now know as the romance because of, among other things, its concern with the everyday and its rejection of the supernatural. And it is also the case that in spite of terminological confusions, the first novelists and their contemporaries generally recognized this. Thus as Day also points out, Samuel Johnson's fourth *Rambler* essay (1750) insists upon the same distinction that twentieth-century critics have seen to be central to any explanation of the newness of the novel:

> The works of fiction, with which the present generation seems more particularly delighted, are such as exhibit life in its true state, diversified only by accidents that daily happen in the world, and influenced by

[14]Michael McKeon, *The Origins of the English Novel 1600–1740*, pp. 2–3.

passions and qualities which are really to be found in conversing with mankind.

This kind of writing may be termed not improperly the comedy of romance, and is to be conducted nearly by the rules of comic poetry. Its province is to bring about natural events by easy means, and to keep up curiosity without the help of wonder: it is therefore precluded from the machines and expedients of the heroic romance, and can neither employ giants to snatch away a lady from the nuptial rites, nor knights to bring her back from captivity; it can neither bewilder its personages in desarts, nor lodge them in imaginary castles.
. . .

The task of our present writers is very different; it requires, together with that learning which is to be gained from books, that experience which can never be attained by solitary diligence, but must arise from general converse, and accurate observation of the living world.

'Life in its true state', 'accurate observation of the living world': the novel has a relationship with ordinary life, with the *detail* of contemporary experience, both social and individual, which sets it apart from the other literary genres.

Once again, however, we need to proceed with care. For as J. Paul Hunter has pointed out:

We do the novel – and not only the early novel – a disservice if we fail to notice, once we have defined the different world from romance that novels represent, how fully it engages the unusual, the uncertain, and the unexplainable. When we admit such concerns as part of the novel's territory, it also allows us to see more clearly where the novel came from, for the novel is only the most successful of a series of attempts to satisfy, in a context of scientific order, the itch for news and new things that are strange and surprising.

A closely related feature involves the novel's engagement with taboos. Operating from a position of semi-respectability from the start, the novel has always been self-conscious about appearing to be pornographic, erotic, or obscene.[15]

That contrast and tension between 'semi-respectability' and 'pornographic, erotic, [and] obscene' is, if not a standard feature of the novel, certainly one of its recurrent features: the novel typically presents us not just with the ordinary, but with the extraordinary, the hidden, the

---

[15] J. Paul Hunter, *Before Novels*, p. 35.

repressed which are to be found hidden in, or behind, the ordinary, the conventional, and the everyday.

In his excellent introductory book on the novel the critic Arnold Kettle suggests that most novelists show a bias towards either 'life' or 'pattern' in their approach to writing: towards, in other words, either the aim to convey the vividness, the particularized sensations and experiences of living, or that of conveying some interpretation of the significance of life. According to Kettle the novelist who starts with pattern often tries to 'inject' life into it, while the novelist who starts with life tries to make a pattern emerge out of it. He relates these two very general tendencies to, on the one hand, such sources and influences as the parables of the Bible, the morality plays of the Middle Ages, and the sermons which common people listened to every Sunday ('pattern'), and on the other hand to the seventeenth- and early eighteenth-century prose journalism and pamphleteering of such as Thomas Nashe and Daniel Defoe (who, we should remember, was a political journalist before he wrote prose fiction). As Kettle says of such writers, the germ of their books is never an idea, never an abstract concept; they are 'less consciously concerned with the moral significance of life than with its surface texture'.[16] Michael McKeon makes a similar point, arguing that in tracing the emergence of the early modern novel we detect a recurrent formal tension between 'what might be called the individual life and the overarching pattern'.[17]

If we remember that the novel is essentially a telling – a narrative – we can perhaps see why there are these lines of development in the novel, not crudely or obviously present of course, but recognizable still today in the difference between a novel like Kingsley Amis's *Lucky Jim* (1954) and William Golding's *Lord of the Flies* (1954). When we tell someone a story, whether it's about a traffic accident in which we were involved or the history of the boy scout movement, we normally have to make a decision about whether we will concentrate most upon the imparting of 'facts' in order to make a point, or the conveying of feelings and experiences. Thus we are likely to deliver a different sort of narrative to the police officer from that recounted the next day over lunch to friends. In her essay 'Phases of Fiction', Virginia Woolf uses terms similar to those to be found in Kettle's later comments in explaining what she sees to be a fundamental tension within the genre of the novel.

[16] Arnold Kettle, *An Introduction to the English Novel* (London, Hutchinson, 2nd edn 1967), volume 1, p. 20.

[17] Michael McKeon, *The Origins of the English Novel 1600–1740*, p. 90.

For the most characteristic qualities of the novel – that it registers the slow growth and development of feeling, that it follows many lives and traces their unions and fortunes over a long stretch of time – are the very qualities that are most incompatible with design and order. It is the gift of style, arrangement, construction, to put us at a distance from the special life and to obliterate its features; while it is the gift of the novel to bring us into close touch with life. The two powers fight if they are brought into combination. The most complete novelist must be the novelist who can balance the two powers so that the one enhances the other.[18]

It is perhaps the mark of the greatest novels that with them it is most difficult to say whether or not the novelist has tried more to convey what Kettle calls 'life' and 'pattern' and Woolf calls 'design and order' and 'life', for both seem so consummately present.

The emergence of the novel

There are two dangers involved in discussing 'the emergence of the novel' which need to be confronted straight away. The first is that the phrase conjures up a sort of hatching process in which an infant species emerges, small and weak perhaps, but with its whole future development laid out ahead of it. The second and related danger is that what we, retrospectively, see as an emergent species was actually many different things which are related only to the extent that they have all contributed to a later and more well-defined genre. In a rewarding but difficult discussion Michael McKeon takes the term 'the novel' and argues that it has to be understood

> as what Marx calls a 'simple abstraction,' a deceptively monolithic category that encloses a complex historical process. It attains its modern, 'institutional' stability and coherence at this time because of its unrivaled power both to formulate, and to explain, a set of problems that are central to early modern experience.[19]

These problems, according to McKeon, encompass both generic and social categories: questions about how to tell the truth in narrative

---

[18] Virginia Woolf, 'Phases of Fiction' (first publ. 1929, repr. in Virginia Woolf, *Collected Essays* vol. 2, London, Hogarth Press, 1966), p. 101.

[19] Michael McKeon, *The Origins of the English Novel 1600–1740*, p. 20.

('questions of truth'), and questions about 'how the external social order is related to the internal, moral state of its members' ('questions of virtue'). McKeon sees a double process in the emergence of the novel: on the one hand, the change from what he calls 'romance idealism' or 'a dependence on received authorities and a priori traditions' to a 'naive empiricism' based upon a more empirical epistemology; on the other hand the change from 'a relatively stratified social order supported by a reigning world view' that McKeon calls 'aristocratic ideology' to what he refers to as a more 'progressive ideology' which is itself later challenged by a 'conservative ideology'.[20]

I have already indicated that the issue of 'the emergence of the novel' has involved much controversy in recent years, and McKeon's arguments are by no means universally accepted. But most of those concerned to map and to explain the development of the modern novel are agreed in singling out certain factors crucial to this development. Of these the following four are perhaps the most important.

(i) The rise of *literacy*. The novel is essentially a written form, unlike poetry which exists for centuries prior to the development of writing, and still flourishes in oral cultures today. There have been cases of illiterate people gathering to hear novels read – part of Dickens's audience was of this sort, and during the Victorian period the habit of reading aloud within the family was much more widespread than it is today. But the novel is typically *written* by one individual in private and *read silently* by another individual who has no personal relationship with the author. Michael McKeon points out that the development of literacy helps to encourage the growth of 'empirical attitudes' which in their turn encourage 'a more skeptical approach to the authenticity of saints' relics and a more rationalistic interpretation of the figurative status of the Eucharist'.[21] Equally if not more important, I suspect, is the fact that silent reading is private and non-collective and thus helps the novel to mediate and express the new individualism of early capitalism while at the same time allowing easier access to the fantasy life. It is easier to fantasize alone with a book than as a member of a theatre audience.

It is surely not accidental that there was a rapid growth of literacy in the English-speaking world from 1600 to 1800, and that whereas in 1600 it is estimated that about 25% of adult males in England and Wales could

---

[20] Michael McKeon, *The Origins of the English Novel 1600–1740*, p. 21.

[21] Michael McKeon, *The Origins of the English Novel 1600–1740*, p. 35.

read, by 1800 the figure is probably between 60% and 70%.[22] Although the figures for female literacy lag behind those for male literacy, more and more women too could read during this period.

(ii) *Printing*: the modern novel is the child of the printing press, which alone can produce the vast numbers of copies needed to satisfy a literate public at a price that they can afford. Michael McKeon points out that a number of different factors conspire to assist the development of printing – technological, social, and legal:

> By the close of the seventeenth century England had become a major producer of paper, type, and various kinds of publication, a development that could not have occurred without the abolition of protectionist printing legislation. By the middle of the eighteenth, licensing laws had been replaced by copyright laws; the commodification of the book market as a mass-production industry had become organized around the mediating figure of the bookseller; and the idea that a writer should get paid for his work was gaining currency on the strength of the perception that there was a growing mass of consumers willing to foot the bill. On the other hand, these new consumers of print were also produced by print – most obviously by the early modern 'educational revolution,' which received a great stimulus from the invention of the press and played the central role in raising the levels of literacy in England.[23]

J. Paul Hunter sums up as follows:

> From the beginning, moreover, novels were artifacts of the world of print. They had been conceived that way, and they lent themselves readily to the physical, social, and psychological circumstances that affected potential readers of printed books.[24]

One of the things that print brought with it was a change in the relationship between reader and writer: print allows for a more impersonal, even anonymous writing – but one that, paradoxically, by cutting the reader off from a known writer, allows him or her to feel that the reading of a novel is a personal, even intimate experience. As I have said, the novel is read *in private* by an individual. Experiencing a novel is thus a much less collective and public matter than experiencing a performed play can be,

---

[22] Figures from J. Paul Hunter, *Before Novels*, pp. 65–6.

[23] Michael McKeon, *The Origins of the English Novel 1600–1740*, p. 51.

[24] J. Paul Hunter, *Before Novels*, p. 41.

where we are very conscious of how the rest of the audience is reacting. J. Paul Hunter has noted the novel's liking for the confessional and the exhibitionistic;[25] the problem-pages of popular magazines will remind us that confession and self-exposure are often easier through the impersonal medium of print.

Print did not succeed oral delivery in one fell swoop; David Margolies points out that the manuscript circulation of written work represented a sort of in-between stage, and 'provided a model for the literary relationships of the early novelists'.[26] Margolies notes that writing for anonymous readers presented writers with a new set of problems, and he quotes from Austen Saker's *Narbonus* (1580):

> and he must write well that shall please all minds: but he that planteth trees in a Forest, knoweth not how many shall taste the Fruit, and he that soweth in his garden divers Seeds, knoweth not who shall eat of his Sallets. He that planteth a Vine, knoweth not who shall taste his Wine: and he that putteth any thing in Print, must think that all will peruse it: If then amongst many blossoms, some prove blasts, no marvel if amongst many Readers, some prove Riders.[27]

Compare this to the opening words of Chapter 1 of Henry Fielding's *Tom Jones* (1749):

> An author ought to consider himself, not as a gentleman who gives a private or eleemosynary [i.e. charitable] treat, but rather as one who keeps a public ordinary [i.e. a public house], at which all persons are welcome for their money.

The relationship between the writer of a novel and its reader is, then, typically impersonal and commercial, rather than one based on personal acquaintance and friendship. This is such a familiar state of affairs for us today that we pay no attention to it: for us, it is a matter of note if we *do* know the author of a book we are reading. But we should remember that this situation is historically new and is associated with the development of printing.

---

[25] J. Paul Hunter, *Before Novels*, p. 37.

[26] David Margolies, *Novel and Society in Elizabethan England* (Beckenham, Croom Helm, 1985), p. 23.

[27] David Margolies, *Novel and Society*, p. 24.

This is not to say that novelists – Fielding included – were not to become adept at convincing their readers that they were in intimate contact with a person interested in them as individuals. As J. Paul Hunter argues:

> [The novel's] rejection of familiar plots, conventional mythic settings, and recognizable character types represents a recognition that its future lies in developing a new kind of relationship with new combinations of readers.
>
> . . .
>
> One audience-related feature involves the novel's tendency both to probe and promote loneliness and solitariness, rather ironic in view of the novel's expressed design to portray people in their societal context. If the tendency to expose secrets of personal life points one way, the tendency to enclose the self, to treat the self as somehow inviolable and insistently separate in spite of the publication of secrets once thought too personal to articulate, points the other.[28]

Hunter makes the astute point that this may well be why the novel, although born in another time, is so well placed to represent some central experiences that are peculiar to the twentieth century: a concern with subjectivity, isolation, and loneliness.[29]

As David Margolies claims, writers did not necessarily learn the new ways of writing that we associate with print without difficulty; early printed works often show signs of a still-flourishing oral tradition, and J. Paul Hunter has argued that the first modern novels

> had their relevant contexts – ultimately even their origins – in a culture that was partly oral and partly written, where functions traditionally performed in communal and family rituals and by oral tradition more and more fell to the impersonal processes of print.[30]

We should not look upon this as a disadvantage: the novel owes much of its richness to the importation of techniques and perspectives from oral delivery. From its earliest days the novel has been able to make readers think that they are hearing a voice when they are actually reading print, and the gains of verisimilitude and dramatic immediacy that this has

---

[28] J. Paul Hunter, *Before Novels*, p. 39.

[29] J. Paul Hunter, *Before Novels*, pp. 41–2.

[30] J. Paul Hunter, *Before Novels*, p. xvii.

meant are considerable. Nor is the influence of oral traditions on the novel limited to the eighteenth century; one of the striking characteristics of many novels written in the twentieth century by writers from the working class of industrialized societies and by writers from newly independent ex-colonies is that oral rhythms and structures are taken from a rich heritage of non-literary traditions. Take the opening of Jack Common's semi-autobiographical novel, *Kiddar's Luck* (1951):

> She was a fool, of course, my mother. Her mother said so: 'Bella is a fool, I'm afraid, a weak fool. Here she is marrying a common workman, one who drinks and is not a good Christian. She will never know happiness now.' You would think the old lady was great shakes herself to hear her. And she was in her way. Not that she had any money ever, but she made poverty respectable.

The intimate relationship established here between writer and reader feeds off a living tradition of oral storytelling at the same time that it exploits the potentialities inherent in the privacy and anonymity central to the reader–writer relationship associated with the novel.

(iii) A *market economy*. The 'sociology of the novel' is based very much upon a market relationship between author and reader mediated through publishers. In contrast to earlier methods of financing publication or supporting authors such as patronage (a rich patron would support a writer while a book was being written) or subscription (rich potential readers would subscribe money to support a writer in order that a particular work might be written), a market economy increases the relative freedom and isolation of the writer and decreases his or her immediate dependence upon *particular* individuals, groups, or interests (although, of course, it makes him or her *more* dependent upon publishers and, especially, sales). The growth of a market economy is an integral aspect of the rise of capitalism – the socio-economic system which had displaced feudalism in Britain by the eighteenth century. In his book *The Rise of the Novel* (1957) Ian Watt has argued for a close relationship between the rise of the novel and the rise of the middle class – the class most involved in the triumph of capitalism in Britain. In different ways literacy, printing and a market economy can all be related to the growing dominance of capitalism in the period during which the novel emerges.

Michael McKeon draws attention to the debate in Cervantes's *Don Quixote* between Don Quixote and Sancho Panza about whether the latter should receive 'favours' or 'wages', suggesting that the relationship between these two characters thus 'mediates the historical transition from feudalism to capitalism'.[31] J. Paul Hunter has gone further and has suggested that crucial to the development of the new genre in England is the confrontation between 'a Protestant, capitalistic, imperial, insecure, restless, bold, and self-conscious culture', and 'a constrictive, authoritarian, hierarchical, and too-neatly-sorted past'.[32]

It is certainly the case that the novel from its earliest days is concerned with conflict, and its portrayal of confrontations between individuals very often has a representative quality, pointing in the direction of larger social, historical, or cultural rivalries. This is significant not just in terms of the content of novels, but (and this is of course not a separate issue) their readership. Novels appealed to groups of individuals who were emerging into new prominence and increased power: women, and young people, for example. Hunter points out that from its earliest days the novel has been particularly concerned with 'the crises of the decisive moments in adolescence and early adulthood'.[33] A novel such as Jane Rule's splendid *Memory Board* (1987), which focuses on characters who are old and tackling senility, has few parallels before the present century.

Probably the most crucial developments (for the creation of the novel) brought about by capitalism constitute my fourth point.

(iv) The rise of *individualism and secularism*. Ian Watt sees as typical of the novel that it includes 'individualization of . . . characters and . . . the detailed presentation of their environment'. Unlike many of the narratives that precede it the novel does not just present us with 'type' characters; we are interested in Tom Jones, David Copperfield, Maggie Tulliver and Paul Morel as distinct individuals with personal qualities and idiosyncrasies. As Ralph Fox puts it, somewhat tendentiously:

> The novel deals with the individual, it is the epic of the struggle of the individual against society, against nature, and it could only develop in a society where the balance between man and society was lost, where man was at war with his fellows or with nature. Such a society is capitalist society.[34]

---

[31] Michael McKeon, *The Origins of the English Novel 1600–1740*, p. 283.

[32] J. Paul Hunter, *Before Novels*, p. 7.

[33] J. Paul Hunter, *Before Novels*, p. 43.

[34] Ralph Fox, *The Novel and the People* (London, Lawrence & Wishart, 1979; first publ. 1937), p. 44.

Fox exemplifies his argument by contrasting the *Odyssey* with *Robinson Crusoe* (1719), pointing out that whereas Odysseus lives in a society without history and knows that his fate is in the hands of the gods, Robinson Crusoe is prepared to make his own history. Moreover, whereas Odysseus's efforts are directed towards returning *home*, for Crusoe it is the outward and not the homeward trip that is important, he is 'the man who challenges nature and wins'. Crusoe is thus, for Fox, representative of that spirit of expansionism, self-reliance and experimentation that characterizes early capitalist man.

It certainly seems to be the case that the new spirit that accompanies the early development of capitalism infuses the emerging novel. Along with a stress on individualism goes, too, a growing concern with the inner self, the private life, subjective experience. As the individual *feels* him or herself an individual, rather than a member of a static feudal community with duties and characteristics which are endowed at birth, then he or she starts more to think in terms of having certain purely personal rather than merely communal interests. And this gives the individual something to *hide*. Without wishing to oversimplify an extremely complex and far from uniform historical development we can say that in a certain sense the private life as we know it today is born with capitalist society, and that the novel both responds and contributes to this development.

It doubtless seems very odd to many readers to claim that people have not always felt themselves to be individuals in the way that modern men and women from developed societies do, and indeed this assertion has been challenged by a number of recent writers, including Margaret Anne Doody in her *The True Story of the Novel*. Other commentators have detected Eurocentric and even racist attitudes behind the assumption that those from non-literate cultures are not 'individuals' in the way that those from modern developed cultures are.

However fictional works do offer strong evidence to support the view that people from non-literate cultures perceive themselves and their individuality rather differently from the way in which those from literate cultures do.[35] In an important study, *Form, Individuality and the Novel*, Clemens Lugowski examines a number of much earlier fictional works dating from the sixteenth and early seventeenth centuries, including the anonymous *Eine schöne Historie von den vier Heymonskindern*, Georg Wickram's *Amadis*, the stories known as *Thyl Ulenspiegel*, and the

---

[35] A classic case for this view is made by A.R. Luria in his book *Cognitive Development* (Cambridge, Mass., Harvard University Press, 1976).

novellas of the *Decameron*. He concludes that these works contain nothing that might be referred to as an 'individual'.

One of the things that the novel both reflects and helps to establish – in the form which it assumes from Defoe onwards – is a fundamental belief that what a person is goes beyond what that person says or does. Prior to this, Lugowski argues, a view of fictional characters as more components of a whole than as genuinely autonomous individuals means that genuinely individual characters as we understand them do not exist.[36] Lugowski's argument assumes that the ability to distinguish oneself from the world in which one lives is not universal, and that its gradual emergence encourages and takes support from the emergence of the novel in its modern form:

> [W]hen the autobiographer explicitly *wishes* to describe his own life, and yet then describes the world outside, then what this means is that his life really *is* the world . . .[37]

This point can be linked to one made earlier: the emergence of the modern novel is linked to *conflict* and social tension. For, as Lugowski says, it is when human beings begin to question their 'position in and profound commitment to the world' that an individuality that involves seeing oneself as separate from the world can emerge. If the world is taken for granted, how can we define ourselves by seeing ourselves as, in some sense, separate from it?[38] To summarize: the fact that the novel is closely associated with the rise of individualism and the fact that the novel typically deals with conflict are related.

In a similar vein, M.M. Bakhtin says of the work of Rabelais (which he much admires), that in it 'life has absolutely no *individual* aspect. A human being is completely external.' He continues:

> For indeed, there is not a single instance in the entire expanse of Rabelais' huge novel where we are shown what a character is thinking,

---

[36] Clemens Lugowski, *Form, Individuality and the Novel* (trans.) John Dixon Halliday (Oxford, Polity Press, 1990), p. 83.

[37] Clemens Lugowski, *Form, Individuality and the Novel*, p. 155.

[38] Clemens Lugowski, *Form, Individuality and the Novel*, p. 176.

what he is experiencing, his internal dialogue. In this sense there is in Rabelais' novel no world of interiority.[39]

Of all literary genres the novel most consummately unites an exploration of the subjective and the social, of the private and the collective. The lyric poem has offered us highly personal statements of individual feeling and cerebration, but without the ability to contextualize these in the manner of the novel's detailed depiction of the complex relationships between subjective feeling and history and society. Drama has provided us with human beings interacting within the pressures of a given society at a particular time, but its most sophisticated verbal portrayals of subjective experience – Hamlet's soliloquies, for instance – are technically crude in comparison to what the novel can achieve in this direction. Even the simplicity of that opening sentence from James's 'The Lesson of the Master' gives us something that is beyond the scope of the lyric poem and the drama (or film). It is this combination of the broadest social and historical sweep with the most acute and penetrating visions of the hidden, private life, *and their interconnections*, that is characteristic of the novel and at the heart of its power and continuing life. And paradoxically, although the novel is both written and read in private, it relies upon a highly organized society and industry to produce and circulate it. Even in its sociology it combines the personal and the social, that combination that is at the heart of its aesthetic.

This is not to say that the novel is able critically and analytically to portray all human organizations and groups as well as it can represent and explore the individual. Brook Thomas has, indeed, suggested that 'the generic demands of the novel seem to guarantee that novelistic representations of corporations will lag behind the way in which other forms of discourse have been able to respond to the rise of corporations,'[40] implying that the novel is generically unfitted to portray and examine human beings in certain of their collectivities.

When we say that *secularism* is crucial to the development of the novel this is not to imply that a novel may not explore religious themes or be underpinned by a religious imperative. But it does mean that the modern novel emerges in a world in which people were more and more

---

[39] M.M. Bakhtin, 'Forms of Time and Chronotype in the Novel', in M.M. Bakhtin, *The Dialogic Imagination: Four Essays* (Austin, University of Texas Press, 1981), p. 239.

[40] Brook Thomas, *The New Historicism: And Other Old-fashioned Topics* (Princeton, Princeton University Press, 1991), p. 150.

likely to try to find non-supernatural explanations for the problems which they faced, and that this cast of mind is reflected in the novel. Even when a novelist wishes to make a religious point, he or she is typically encouraged by the genre to set it in a context of secular explanation. Thus when a character in David Lodge's *Small World* (1984) prays for help when faced with the imminent task of delivering a conference paper which he has not yet written, his prayer is answered by a sudden outbreak of Legionnaire's Disease which requires that the conference be abandoned, and the reader is allowed to read this as either divine intervention or a fortunate coincidence. It is true that there are novels which encourage the reader to imagine supernatural occurrences and fantastic worlds different from our own (the gothic novel, science fiction, magic realism), but it is striking how consistently such novels hedge their bets by allowing for non-supernatural explanations or interpretations of what happens. It is for this reason, among others, that modern readers tend to find the novels of Ann Radcliffe more acceptable than they do her near-contemporary William Beckford's *Vathek* (1787), in which the supernatural interventions allow for no natural explanation.

The early association of the novel with town rather than country life is also significant. There are novels set in the country of course, but from its earliest days the novel appears to have had a special relationship with town life, and both the readers and the writers of novels were more likely to be town-dwellers than country-dwellers in the eighteenth century. If we look at what has a fair claim to be one of the first modern novels – Daniel Defoe's *Moll Flanders* (1722) – we can see that the town and the novel form have much in common. Both involve large numbers of people leading interdependent lives, influencing and relying upon one another, but each possessing, nevertheless, a core of private thoughts and personal goals. Modern readers may need to be reminded that the growth of large towns and of urban life usher in new forms of loneliness and privacy at the same time that they depend upon and bear witness to collective planning and human co-operation.

Consider the following passage from *Moll Flanders*, which concludes the scene in which Moll robs a young child of a necklace:

Here, I say, the Devil put me upon killing the Child in the dark Alley, that it might not Cry; but the very thought frighted me so that I was ready to drop down, but I turn'd the Child about and bade it go back again, for that was not its way home; the Child said so she would, and I went thro' into *Bartholomew Close*, and then turn'd round to another Passage that goes into *Long-lane*, so away into *Charterhouse-Yard* and out into *St. John's-Street*, then crossing into *Smithfield*, went down *Chick-lane* and into *Field-lane* to *Holbourne-bridge*, when mixing with the Crowd of People usually passing there, it was not possible to have been found out; and thus I enterpriz'd my second Sally into the World.

How well the world of London and the world we now associate with the novel fit! Rapid mobility in a well-known world – the everyday world with which so many readers are familiar – alongside a sense of one's private self which, even in the midst of crowds, is one's own alone (and the reader's!).

It is interesting, however, to note that from its earliest days the novel seems often to split not just between novels where the author starts with 'life' and those in which the author starts with 'pattern' – to use Arnold Kettle's terms – but between novels in which the author is more interested in the public world and novels in which the author is more interested in private life. Again, it is only the very greatest novels that seem to combine the two such that we feel no sense of subordination of either. Henry Fielding's *Tom Jones* (1749) and Samuel Richardson's *Clarissa* (1747–8) can be taken as representative here, with the former's greater interest in a masculine, public life of movement, action and life in the larger social world in sharp contrast to the latter's concentration upon a feminine, more inward life of feeling, personal relationships, and personal moral decision. This is not to deny that Fielding's novel includes a concern with feeling and moral duty and Richardson's with the larger social context of the inner world. Nonetheless there is a significant difference between the two writers' foci of interest, a difference which is perceptively brought out in a conversation reported by James Boswell in his *The Life of Samuel Johnson*, which was first published in 1791. Talking to Johnson and Thomas Erskine, Boswell expressed surprise at Johnson's calling Fielding a 'blockhead' and 'a barren rascal', and continued:

'Will you not allow, Sir, that he draws very natural pictures of human life?' JOHNSON. 'Why, Sir, it is of very low life. Richardson used to say, that had he not known who Fielding was, he should have believed he was an ostler. Sir, there is more knowledge of the heart in one letter of Richardson's, than in all *Tom Jones*. . . . ERSKINE. 'Surely, Sir, Richardson is very tedious.' JOHNSON. 'Why, Sir, if you were to read Richardson for the story, your impatience would be so much fretted that you would hang yourself. But you must read him for the sentiment, and consider the story as only giving occasion to the sentiment.' [Anecdote from 1772]

We do not need to accept Johnson's value judgements to feel the force of the distinction made here, between a primary interest in drawing very natural pictures of life – albeit low life – and that of revealing knowledge of the human heart.

I talked earlier about Fielding's interest in the masculine and Richardson's in the feminine sphere, and I should make it clear that this is a historical rather than a universal judgement; 'masculine' and 'feminine' as they were defined by eighteenth-century English society, in which men were able to live a public as well as a domestic life, a public life of travel, work, exploration and adventure. The relationship of women to the novel is a very important one, however, which has already been touched upon. This importance is both as writers and as readers. Although the early novelists in England and elsewhere were predominantly men up until the late eighteenth century, women soon formed a substantial and, on occasions, a dominant section of the reading public. This, of course, because of the market economy on which novel production depended, had an appreciable effect upon the sort of novels that were written, and it is impossible to imagine Richardson's novels having been written as they were without a substantial female reading public.

But by the nineteenth century women were a dominant element not just as readers but as writers of the novel. What male novelist is there then to compare with Jane Austen, Emily Brontë and George Eliot in England, apart from Charles Dickens? It is arguable that the novel's success in exploring the private world, the subjective self, could never have been accomplished without the contribution to the genre of that introspective self-knowledge and sensitive perception of interpersonal relations that women's domestic imprisonment had trained them to be so expert in. No man could have written Jane Austen's novels.

Perhaps we can add another point to our list of the novel's achievements: no other literary genre has enriched itself so fully and so impartially from the culture and experience of both sexes. (Which is not, of course, to say that the novel is innocent of the blinkered vision and perceptual shortcomings of women or, especially, of men.)

The international life of the novel: form and culture

From its first appearance in the world the novel is associated with movement and travel – and, in a more general sense, with *mobility*. In the passage I quote above from *Moll Flanders* we are reminded how important being able to move from place to place was for Moll, but we should remember, too, that for her *geographical* movement was closely linked to her desire for *social* movement or advance. It is certainly arguable that this stress upon mobility is one of the points at which the socio-historical conditions of the novel's emergence and the social tensions which contribute to its birth can be seen transposed into generic and formal features. The very name 'novel', of course, directs our attention to what is new, and we typical discover the new by changing our location and by seeing either new things or old and familiar things from an unusual perspective. Peter Burke has suggested that travel is one of the factors which can help to explain the emergence of what he refers to as 'the autobiographical habit' in the sixteenth century. As he reminds us, this 'was the age not only of the rise of the autobiography or journal but also of fictional narratives in the first-person story, such as the picaresque novel in Spain or the sonnet-sequences of Sidney, Shakespeare and others', and he proposes three likely reasons for this development. First, the fact that this is the age in which print becomes a part of everyday life, which means that the diffusion of printed models can create a new or sharper sense of self. Second, urbanization, because the alternative styles of life offered by the city encourage a sense of individual choice. And, finally, travel, because travel 'encourages self-consciousness by cutting off the individual from his or her community'.[41] While we remain in familiar surroundings we receive more regular confirmation of our identity from those who expect us to continue to be what we have been in the past, but in a strange situation, surrounded by people we do not know, we become aware that to a certain extent we are able to *choose* who we are.

Even that basic component of prose fiction – the linearity of narrative – models progression and movement for us. As the line of prose snakes its way ahead from page to page, always moving in a direction our culture calls 'forward' and always moving away from its point of birth to that last line where it will stop forever,[42] it constantly reminds us not just of the

---

[41] Peter Burke, 'Representations of the Self from Petrarch to Descartes', in Roy Porter (ed.), *Rewriting the Self: Histories From the Renaissance to the Present* (London, Routledge, 1997), p. 22.

[42] 'Always' is perhaps an exaggeration. Some writers have attempted to 'de-linearize' the novel; Brian McHale (*Postmodernist Fiction*, London, Routledge, 1987, p. 193) refers to two of the best-known examples – Julio Cortázar's *Hopscotch* (1963), which invites the reader to choose between two different

individual human life journeying from birth to death but of the typical fictional hero or heroine: progressing ever forward, continually encountering new blank pages untouched by the pen of experience. As Michael McKeon puts it, by arranging events in a certain linear order, that 'progressive narrative' which is integral to the emerging novel

> automatically invests personality with the distinct moral qualities that are implied in the condition of being either in decline or on the rise. These qualities imply, in turn, an explanatory rationale whose force is unmistakable. On one side are the extravagant and licentious older sons of nobility, sunk in decay and corruption; on the other, the industrious and virtuous younger sons or tradesmen, hard at work in their honest and quasi-Calvinist callings.[43]

McKeon's argument links a specific type of narrative with a particular sort of social mobility, but these two forms of directed movement are also and typically associated with geographical progression and movement in the novel. We should remember, too, that, although the reader of the novel is very much the captive of its linear progression, at the same time he or she can have recourse to various strategies of resistance: pausing, re-reading, scanning the pages more quickly or more slowly and carefully. This control of the way in which the text is 'consumed' is not open to the spectator or audience of the performance arts. Like the hero or heroine of a novel, the reader must follow a single line of development forward, but can choose exactly how this line is followed.

A play may well conform to the unities of time, place and action, but for us the novel is rarely so confined. There are short stories which accept such restriction, but even in these cases it is rare to find a story which deals only with a group of people who have all been living in the same place for years and years. The stranger, the intruder, the traveller – whether the terrain crossed is geographical or social – are all familiar inhabitants of this 'novel' genre.

There is, then, a relationship of mutual dependency between the novel's concern with the new, its questioning of laws and conventions, its curiosity about the strange and unusual, and its stress on geographical

---

routes through its text, and B.S. Johnson's *The Unfortunates* (1969), which was (and is) published in a box of 27 stapled gatherings which are to be shuffled and read in the orders that result. In both cases, one should note, any individual reading is still a linear progression from beginning to end through a finite text.

[43] Michael McKeon, *The Origins of the English Novel 1600–1740*, p. 220.

mobility. But 'geographical mobility' is a rather suspiciously abstract and bloodless term. When people move around the world they meet and interact with other people with whom they may have a range of different sorts of relationship: collaborative, subservient, antagonistic, competitive, or exploitative. The travel experiences of a tourist are not those of a refugee; those of a soldier are not those of an ambassador. The novel in the sense that I have suggested we adopt emerged in a world in which relationships between peoples were certainly not of an exclusively collaborative or equal type; by the end of the seventeenth century Europe was already some way into those processes of exploitation and repression which we know as colonialism and imperialism.

Firdous Azim has pointed out that from the moment of its birth the novel enjoyed an 'imperialist heritage', one which is visible both in its themes and also in the problematic nature of its status: '[t]he novel becomes a form dealing with adventures in far-off lands, Oroonoko and Robinson Crusoe on the one hand, and the sexual adventures of bold and courageous women at home – Zara, Rivella, Moll Flanders and Roxana, on the other'.[44] (We might add that Moll Flanders's adventures are not just 'at home': she also travels to America, where a significant portion of Defoe's novel takes place.)

'Imperialist heritage' may sound somewhat tendentious, but it is nevertheless a fact that the novel emerges in western Europe at a time when this part of the world is in the early stages of a long process of imposing its will and influence – by both peaceful and non-peaceful means – on other peoples. Thus the master-slave relationship in, for example, *Robinson Crusoe* has to be seen as more than just an interesting part of the story, an element in the work's content. Defoe's hero explores his own individuality in a number of ways, but one that is of defining importance to this novel, as well as to the experience of its author and his or her fellow citizens, is the imposition of one's individual and collective will on the inhabitants of other lands. Crusoe discovers who he is – *makes* his identity – in part by establishing who Friday is and what their relationship is to be.

A number of recent studies of the novel which, either with or without the approval of their authors, have been dubbed 'postcolonialist', have suggested that it is not just during the period of its emergence and generic consolidation that the history of the novel and the history of imperialism are related in complex and significant ways. Judie Newman's book *The*

---

[44]Firdous Azim, *The Colonial Rise of the Novel* (London, Routledge, 1993), p. 21.

Ballistic Bard: Postcolonial Fictions, looks at a range of recent novels which have attempted to confront, expose, or reject this perceived 'imperialist heritage'. One of the most interesting chapters in the book focuses on Jean Rhys's novel *Wide Sargasso Sea* (1966) and its attempted settling of accounts with Charlotte Brontë's *Jane Eyre* (1847). Brontë's novel has of course already served as a mine of hidden or repressed cultural information for feminists: the very influential study *The Madwoman in the Attic*[45] takes its title from the situation of Rochester's concealed and incarcerated first wife, and the book presents her situation as in many ways emblematic of a much wider repression and denial of female oppression – and not just in Brontë's culture. This first wife and her hidden imprisonment impact in crucial ways not just on Rochester but also on Jane – a reminder that the identity and experience of women left in the 'home country' may be fundamentally influenced by what men do in the colonies or in 'savage' lands. (Compare the way in which in Joseph Conrad's *Heart of Darkness* the life of Kurtz's 'Intended' ends up by being utterly dominated by Kurtz's own 'attic': the secret of his experiences in Africa.)

This leads to the first of two questions which I want to pose about the very genre of the novel: to what extent did the novel's intimate and productive relationship with geographical mobility impose an essentially masculine identity upon it in times when such mobility was very much a male preserve? Must the novel lock up women in the attic or a room of their own in order to highlight male freedom of movement? Can a novel present female experience only as Richardson does in *Pamela* and *Clarissa* – through the depiction of incarceration?

For Jean Rhys that hidden presence in Rochester's attic had a rather special significance: as Judie Newman puts it, for her 'the vital point was that Bertha was West Indian, a white Creole from Jamaica'. Rhys's interest was doubtless related to her own problematic upbringing in the West Indies and even more problematic travel to England, but her taking what has generally been experienced as a marginal or peripheral element in Brontë's novel and making it a central issue in her own intertextual reworking of *Jane Eyre* has had the effect of allowing modern readers to approach the earlier work from a revealingly different perspective.

Newman points out that Charlotte Brontë's portrayal of Bertha is designed to obliterate all sympathy for her.

---

[45]Sandra M. Gilbert and Susan Gubar, *The Madwoman in the Attic: The Woman Writer and the Nineteenth-century Literary Imagination* (New Haven, Conn., Yale University Press, 1979).

[Bertha] is described . . . in terms which appeal to both racial and sexual prejudices. Her hereditary madness, which is supposedly accelerated by sexual excess, clearly reflects Victorian syphilophobia. (The nineteenth century had shifted the point of origin of syphilis to Africa.) Brontë's Bertha has 'a discoloured face', 'a savage face' with 'fearful blackened inflation' of the features: 'the lips were swelled and dark'. Successively described as a demon, a witch, a vampire, a beast, a hyena, and even an Indian Messalina, Bertha unites in one person all the available pejorative stereotypes.

Newman draws attention to the description by (Brontë's) Rochester of Jamaica as 'hell', its sounds and scenery those of 'the bottomless pit'. She also notes that when Rochester contemplates suicide he is saved by a 'wind fresh from Europe', and she reminds us that 'Penny Boumelha has observed that in *Jane Eyre* all the money comes from colonial exploitation' and that 'Jane herself gains her financial independence as a result of a legacy from an uncle in Madeira who is connected to the same firm which Mr. Mason, Bertha's brother, represents in Jamaica'. Finally, Newman quotes again from Penny Boumelha, and provides us with a phrase very similar to one already taken from Firdous Azim. For Boumelha, Rhys's vindication of Brontë's 'madwoman' serves to display and to criticize 'the legacy of imperialism concealed in the heart of every English gentleman's castle'.[46]

Now clearly this legacy, like many other legacies, is not divided equally among the heirs. It is no surprise to find that Joseph Conrad, whose entire writing career seems to be intimately related to his experience of imperialist brutality, should be found to have inherited rather a large portion in spite of his non-British origins, and to have agonized about his use of it. A novel such as Conrad's *Lord Jim* (1900) explores the way in which Conrad's eponymous hero's personal qualities are bound up with the age and the culture to which he belongs. Early on in the novel we are taken into the world of the young Jim's fantasizings.

On the lower deck in the babel of two hundred voices he would forget himself, and beforehand live in his mind the sea-life of light literature. He saw himself saving people from sinking ships, cutting away masts in a hurricane, swimming through a surf with a line; or as a lonely

---

[46] Judie Newman, *The Ballistic Bard: Postcolonial Fictions* (London, Arnold, 1995), p. 14. Penny Boumelha's comments come in her article 'Jane Eyre, Jamaica and the Gentleman's House', *Southern Review* 21(2), July 1988, pp. 111–22.

castaway, barefooted and half naked, walking on uncovered reefs in search of shellfish to stave off starvation. He confronted savages on tropical shores, quelled mutinies on the high seas, and in a small boat upon the ocean kept up the hearts of despairing men – always an example of devotion to duty, and as unflinching as a hero in a book.

'Something's up. Come along.'

He leaped to his feet. The boys were streaming up the ladders. Above could be heard a great scurrying about and shouting, and when he got through the hatchway he stood still – as if confounded.[47]

The passage is interesting in as much as it suggests that Conrad's hero, as much as Robinson Crusoe, has defined himself in part at least in terms of his presumed relationship to 'savages' confronted on 'tropical shores'. His inner life reflects a desired public life, one which will be modelled on a pattern of racial superiority and the exerting of national superiority by means of geographical movement. Ironically, as the quoted passage demonstrates, this fantasized inner life of action and decisive movement is matched by – is perhaps the reason for – an outer life that is character-ized at this stage by an inability to act.

At this stage in the development of the novel, in other words, writers are beginning consciously to interrogate that imperialist heritage of which both Azim and Boumelha have written, and to interrogate it as much in terms of what it does to the consciousness of the 'heroic' European as in terms of its effect on the 'savage'.

Now of course not all novelists have experienced imperialism at first-hand in the way in which Conrad did in the Belgian Congo (the experi-ences which contribute directly to his *Heart of Darkness*). But, as the case of Charlotte Brontë demonstrates, larger social and political inequalities and injustices can be reflected and reformed in the consciousness of those who remain in the domestic (both national and familial) sphere.

If the novel was born at a particular time and place and was the child of distinct experiences and circumstances, today it is very much a citizen of the world. All literate cultures appear to possess some literary genre which resembles the novel to a greater or a lesser extent. Is the novel free of the taint of its birth, is it now a neutral form in which any content, any ideological belief, can find a home? Or is it a genre whose very nature has a particular set of values attached to it, one which cannot escape from

---

[47] Joseph Conrad, *Lord Jim* (World's Classics edition, ed. John Batchelor, Oxford, Oxford University Press, 1983), p. 6.

a set of meanings which predate each particular set of words that constitutes every individual novel?

In a sense I have already aired this issue by asking whether the novel's association with geographical and other mobility imposes a certain 'masculinity' upon it, and by referring to Brook Thomas's suggestion (p. 35) that the genre is not well fitted to the portrayal of corporations, thereby implying that it is generically better suited to depicting individuals than collectivities. But the question has other aspects. Is the novel's formative association with capitalistic expansion, with individualism, with a secular world-view, and with imperialism, one which defines the genre as a whole as (for example) non-collective, 'western', and masculinist? (As *Heart of Darkness* reminds us, imperialism was and is staffed by men: women were left in Europe.)

This question needs to be set alongside another issue, one which is concerned not with the inequalities of the world in which the novel emerged but the inequalities of the world today. Except in unrepresentative or marginal cases, novels live a print-based life. Printing involves not just the technical and economic means to produce and circulate books, but also a universal or at least widespread literacy that comes from a highly developed system of public education. In those countries which have recently emerged from colonial rule much of the publishing business is controlled by foreign, often multinational, companies. The achievement of high levels of literacy in such countries has often relied upon education systems built up by colonial or imperialist powers and modelled upon European or North American lines. It is not so very long ago that little of the literature read by the student of English in many newly independent African countries was actually written by Africans. And even today certain qualifying examinations sat in – say – a relatively newly independent African country may be administered from Europe.

Issues such as these have led a number of commentators to ask not, as I put the matter earlier, whether the novel as a genre is 'non-collective, "western", and masculinist', but whether the novel is or has been used as a weapon of ideological propaganda or control by those cultures within which it emerged and from which it allegedly took its defining characteristics. The two questions – or accusations – are of course linked: the implication is clearly that the novel can be used as an ideological or political weapon precisely because those characteristics which distinguish it reflect and even celebrate a pattern of human relationships first identified and acclaimed in the late seventeenth and early eighteenth centuries in western Europe.

What force does this challenge to the novel have? It seems to me to be important to insist upon the fact that it is quite wrong to treat a novel written today in Africa or India as just no more than a European import, much like a Mercedes car shipped in parts and only assembled in, say, India. As a starting point, consider the following words by the Indian novelist Mulk Raj Anand. They compose part of the Introduction to his collection of Indian fairy tales.

> The stories contained in this volume were told me by my mother and my aunts during my childhood. The primary inspiration to retell them, therefore, came from the nostalgic memories of the hour when 'once upon a time' began and when one's eyes closed long before the story had ended.
>
> But I also had in mind the fact that in the old stories of our country lay the only links with our broken tradition. I fancied that only by going back to the form of these stories, told by mother to son, and son to son, could we evolve a new pattern for the contemporary short story. Of course, the modern short story is a highly developed folk tale if it is a folk tale at all. But a revival of the short story form . . . seemed a fit occasion to relate it to its more primitive antecedents which, surprisingly enough, seem to lie in the sources of the sheaf of tales which I have gleaned. At any rate, I must confess that although I have taken in much new psychology into my own writing of the short story, I have always tried to approximate to the technique of the folk tale, and the influence of these fairy stories has always been very deep on my own fiction.
>
> These fairy stories can, therefore, be read not only by children, but by those adults who have not forgotten the child in them. And, however foreign they may seem to non-Indians, in their atmosphere and effect, I offer them here, not as something completely alien to the Western peoples, but as familiar and well known themes to set beside the fairy tales which they have read in their childhoods, because there has been much international traffic in folk lore between India and the West through traders, travellers, gypsies, craftsmen and crusaders, and many of the stories current abroad have their source in the same springs in which these stories have their origin.[48]

The comments are interesting for a number of reasons. First, because they remind us that, just as a tradition of fictional tales emerged and established itself in Europe before literacy became widespread, so too can

---

[48]*Indian Fairy Tales*, retold by Mulk Raj Anand (Bombay, Kotub Publishers, 1946), pp. 7–8.

comparable traditions be found in those cultures and societies which were on the receiving end of colonialism and imperialism. Second, because, as Anand points out, the stories with which he is concerned strike non-Indian readers as both 'foreign' and familiar. Oral prose fiction traditions may all share some common or familial features, but they are also possessed of that which is culture-specific, and thus to the extent that such traditions inform a new tradition of novel writing and reading this new tradition will also have something that is its alone, something native to its own culture. But third, because, as Anand also points out, 'there has been much international traffic in folk lore', and although the emerging novel may reflect the inequalities and injustices of its mother-culture, it is also the resting place for many perspectives which are critical of that mother-culture and its dominant values. Indeed, as the phrase 'dominant values' may remind us, the societies from which the early novel emerged were by no means monolithic; they contained minority views, 'faultlines', oppositions and contradictions.

Anand's collection of fairy tales was published in 1946, in the dying years of British India. (Indian readers of the collection were probably more immediately aware of what the discreet term 'broken tradition' referred to than are present-day European readers.) Anand knew as well as anyone what imperialism and colonialism were capable of, but he also knew that in addition to the crusaders and the traders there were also travellers, gypsies and craftsmen moving from country to country, and that the commerce was not all one-way. The gypsies came originally from India, and were certainly part of western European culture by the eighteenth century, although of course neither a dominant nor a privileged part.

In short, not only can a novelist writing in newly independent lands incorporate powerful elements of native storytelling tradition into his or her work, but the European tradition of the novel is itself not exclusively European.

Where does this leave us in our quest to determine the extent to which the genre of the novel is or is not value-neutral? It seems to me to be arguable that the novel is probably less well suited to the expression of collective or communal experiences and beliefs than are some other forms such as orally delivered poetry or drama. Not only do such performance arts involve a direct contact between artist-performer and audience, thus allowing both for collective response and also collective feedback from audience to creator, but the drama also allows for the dramatized presentation of distinctive voices – often simultaneously. Now the skilled novelist can resist the individualistic and monologic tendencies of the genre, and indeed I have already referred to Mikhail Bakhtin's belief that the novel is naturally dialogic – unlike poetry which is naturally monologic (see p. 18). Moreover, in my comments on the narrative technique used by Lewis Grassic Gibbon in his *A Scots Quair* (see p. 148) I suggest that this technique is designed to help us to experience a working-class collectivity and that it largely succeeds. Nevertheless, it is arguable that the presentation of different and individualized voices, and the representation of such collectivity, are more difficult for the novelist than they are for the dramatist. What is more, a culture whose vital life is more closely associated with oral expression and an oral tradition than with written expression and a written tradition may find more natural expression in the drama or poetic recital than in the printed pages of a book.

Having said all this, however, it seems to me that if we are asking the crude question, 'Is the novel essentially a bourgeois, individualist, and 'western' (or imperialist) genre?', it is perhaps more productive to look at questions of tradition and economic control than at questions of innate generic identity – in other words, to focus more upon differences of culture and on socio-economic issues than on the inherent formal or technical qualities of the genre. Indeed, so far as the latter are concerned I should repeat that although the novel does seem to be generically predisposed in the direction of representing individual experience, especially when that experience is that of a socially and geographically mobile individual, it is also the case that the novel is a genre which seems to survive by permanently revolutionizing itself, in the light of which fact it is dangerous to talk of what the novel cannot do because of its innate generic identity.

But if we turn to the issues of tradition and economic control we do I think see how particular values and perspectives may be associated with the genre. To explore this issue I would like to take two novels by the Kenyan novelist Ngũgĩ Wa Thiong'o: his first published novel *Weep Not, Child* which was published in 1964 and *Devil on the Cross* which was published first in Gĩkũyũ in 1980 and then in the author's own English translation in 1982. In fact, when *Weep Not, Child* was first published its author's name was given as 'James Ngugi'. Both books were published by the publisher Heinemann, now a part of the Reed International conglomerate.

Weep Not, Child is a wonderful first novel and was welcomed as such on its initial publication. But re-read today from the perspective of Ngũgĩ's later work the novel gives the sense of a writer having to adapt to the requirements of a tradition and a publishing situation which are partly alien to him and which restrict his freedom of movement. Take the following short passage from the opening pages of the novel.

> There was only one road that ran right across the land. It was long and broad and shone with black tar, and when you travelled along it on hot days you saw little lakes ahead of you. But when you went near, the lakes vanished, to appear again a little farther ahead. Some people called them the devil's waters because they deceived you and made you more thirsty if your throat was already dry. And the road which ran across the land and was long and broad had no beginning and no end. At least, few people knew of its origin. Only if you followed it it would take you to the big city and leave you there while it went beyond to the unknown, perhaps joining the sea. Who made the road? Rumour had that it came with the white men and some said that it was rebuilt by the Italian prisoners during the Big War that was fought away from here.[49]

It seems clear that this passage attempts to present the reader both with the consciousness of Njoroge, the main character of the novel, and also with a sense of the, as it were, collective consciousness of Njoroge's people, as well as of the immediate history of his land and culture. Such multiple perspectives are not uncommon in the novel, especially in novels which deal with 'suppressed' or oppressed groups whose lives and experiences are sharply different from those of the writer's intended or expected readers. (Again, my comments on Lewis Grassic Gibbon's narrative technique on p. 148 are relevant to the present discussion, not least with regard to both writers' use of the collective 'you' to affirm and represent shared and communal knowledge.) What the passage gives us is a statement of what, because it is universally shared knowledge, is actually unlikely ever to be stated unless an outsider asks the right questions, and so putting it in Njoroge's unspoken but verbalized thoughts is slightly jarring.

But who is 'the reader'? It seems clear that Ngũgĩ was very conscious of the fact that many of the first readers of this novel would know nothing about Kenya or Kenyans, nothing about the history or culture of this part of Africa. And passages such as the one which I have quoted seem in part designed to impart knowledge to such individuals. Writing a novel, in English, for an international publisher, Ngũgĩ must have felt the pressure of his potential readers' ignorance in a way in which a middle-class

---

[49] James Ngugi, *Weep Not, Child* (London, Heinemann Educational, 1972 [first publ. 1964]), pp. 5–6.

European of his age and generation would not have done. And this doubtless helps to explain a certain stiffness at certain points of the narrative of *Weep Not, Child.* Take the following passage.

> Ngotho often wondered if he had really done well by his sons. If he and his generation had failed, he was ready to suffer for it. . . . But whatever Ngotho had been prepared to do to redeem himself in the eyes of his children, he would not be ordered by a son to take an oath. Not that he objected to it in principle. After all, oath-taking as a means of binding a person to a promise was a normal feature of tribal life. But to be given by a son! That would have violated against his standing as a father.[50]

In this passage it seems to me that the sentence 'After all, oath-taking as a means of binding a person to a promise was a normal feature of tribal life' also jars somewhat: it is sandwiched by sentences which are in Free Indirect Discourse, specifically in represented thought (see p. 109), but it cannot itself be read in this way. This is an authorial interpolation, an explanatory detail provided for the benefit of the reader who is not familiar with the 'tribal life' in question. I am reminded of a comment made by E.M. Forster about Charles Dickens's *Bleak House.* Writing about that part of Dickens's novel which is narrated by Esther Summerson, Forster comments that, even when Esther is ostensibly telling the story, Dickens may snatch the pen from her and take notes himself,[51] and I have a sense of something similar happening in Ngũgĩ's first novel.

Now as Forster's comments demonstrate, the problems which Ngũgĩ faces in his first novel are not unknown to European novelists, many of whom seem to be writing against the grain in their early work. Reading Virginia Woolf's *Night and Day* (1919) after having read her later *Mrs Dalloway* (1925) is to experience that same sense of a creative force which is partly constricted by the generic and formal traditions within which it is operating. The solution for Woolf was to remould the novel to the requirements of her creative impulse, and much the same can be said of Ngũgĩ in the light of his later development as a novelist.

---

[50] James Ngugi, *Weep Not, Child*, pp. 83–4.

[51] E.M. Forster, *Aspects of the Novel* (Harmondsworth, Penguin Books, 1962; first publ. 1927), p. 86.

Moreover, just as Woolf was to complain that the gender biases of a patriarchal society established a hierarchy of values in which those particular to or significant for women were devalued, so too the writer in a newly independent ex-colony is often very much aware that his or her own narrative traditions may possess a very low status among an educated elite or a ruling European minority. Those folk tales or fairy stories mentioned by Mulk Raj Anand, proverbs, riddles, jokes – these are hardly possessed of the same 'serious' status as the experiments in narrative strategy of a James Joyce or a Virginia Woolf. It thus requires a certain amount of self-confidence on the part of a writer struggling for acceptance to allow his or her work to be enriched by them. And self-confidence is not just a matter of individual personality; it relates to the status accorded to oneself within a culture, and the status accorded to that culture by other cultures.

By the time of Ngũgĩ's *Devil on the Cross* we feel that this writer has more than obtained this self-confidence. It is neither accidental nor irrelevant to the argument in hand that this is a novel that is thematically very much concerned with the issues of culture independence and cultural subservience. In the author's own English translation words which appeared in English in the original Gĩkũyũ version of the novel are italicized, so that the very text of the novel gives a sense of cultural struggle, as the author's very language is invaded by English words and phrases. One of the characters – Gatuĩria – actually finds it hard to speak in Gĩkũyũ without peppering his speech with English words: ironically enough he is a 'junior research fellow in African culture', and to double the irony he uses English in order to describe his post.

This is a novel in which we feel that the novelist's technical assurance is directly related to his sense of who he is writing for and why. Where culture-specific details have to be provided for European readers they are given in footnotes rather than being smuggled in to the thoughts or speech of the characters: the novel speaks freely from within a culture to those sharing the culture: its dedication 'To all Kenyans struggling against the neo-colonial stage of imperialism' provides convincing evidence concerning the readers for whom Ngũgĩ intended the work. The whole theme of cultural imperialism is foregrounded in the novel, and this allows Ngũgĩ the freedom to use the resources of his own culture without self-consciousness or a concern that this might not be proper or understood. For all that he is presented as a victim of cultural imperialism, Gatuĩria is aware of the process that is oppressing him and he bemoans the loss of an indigenous cultural tradition: 'Our stories, our riddle, our songs, our customs, our traditions, everything about our national heritage has been lost to us'.[52]

Devil on the Cross does not just propose such a case in abstract terms: by incorporating the force and vitality of such indigenous cultural and narrative traditions into its own telling the novel also enacts a process of cultural self-assertion and revolt. It is very difficult to illustrate this by means of selective quotation, because the shifts between more traditional realism and the fantastic, which are mediated through a range of narrative techniques including 'standard' third-person narrative, riddles and sayings, dance-songs and verses, and other forms unfamiliar to European readers such as that of the 'verse-chanting competition', have a cumulative effect which is lost in brief extracts. But the result is to revolutionize the genre within which Ngũgĩ is working and to open the novel to traditions which are certainly 'un-western', collective rather than individualistic, holistic rather than fragmented. It is important to stress than this does *not* mean that the characters of *Devil on the Cross* are presented as mere puppets, tokens to illustrate a thesis. On the contrary, their specificity as individuals, their particularized humanity, is presented *more* rather than less sharply as a result of their being presented in terms of the past and present of a complex and divided culture.

What this example *does* demonstrate, I think, is that the claim that the genre of the novel is fundamentally bourgeois, individualist, and 'western' (or imperialist) rests upon the fallacy that the generic characteristics of 'the novel' are fixed and unchangeable. But what the novel is today it is partly because of its enrichment by writers such as Ngũgĩ. This is not to deny that literary forms such as that of the novel are involved in international processes of economic, political and cultural conflict, repression and domination, nor that the novel may be better suited to the representation of the individual than the collective consciousness, and to processes of linear development than to complexities of simultaneity and reciprocal interaction. Nor is it to deny that a print-based genre is better adapted to the expressive needs of a literate culture than it is to those of an oral one. But the example of a magnificent novel such as *Devil on the Cross* goes to show that the novel can adapt to and incorporate the strengths of rich oral and communal traditions – whether these are those of Kenya or of the Scottish peasantry and industrial working class about whom Lewis Grassic Gibbon writes.

---

[52]Ngũgĩ Wa Thiong'o, *Devil on the Cross* (London, Heinemann Educational, 1987; first publ. in Gĩkũyũ in 1980 and in the author's English translation in 1982), p. 59.

3 Types of Novel

Categorizing novels may seem rather a tedious and mechanical procedure, but it is alas a fact that we can badly misread (and thus fail to extract the value and pleasure from our reading that we could have enjoyed) if we mistake the type of novel that we are reading. As the previously quoted comment from Johnson on Richardson suggests, one can approach a novel in such an inappropriate way, as a result of misperceiving the sort of novel that it is, that one may end up 'so much fretted' that one feels suicidal! Interestingly, Johnson is commenting upon a near contemporary of his in Richardson: the fact that we are reading a novel written in our own lifetime is no guarantee that we will understand the sort of novel that it is. Because the writers of prose fiction typically attempt to present their readers with something 'novel', even a work written by a contemporary of ours may demand to be read in ways with which we are unfamiliar. But the danger of such a mistaken approach probably increases when we read novels from an earlier time, works written for readers with assumptions and expectations which we may not share.

Like most literary works, novels can be categorized in a number of different ways: with reference to when they were written, with reference to their formal or technical features, or with reference to their major themes. Thus a single novel can often be categorized in a number of different ways. What follows is a list of some of the most commonly used categories, with some explanation and discussion of the terms used to denote these categories.

These categorizations should be used as a guide rather than a strait-jacket: no novel is *just* a picaresque novel or a *bildungsroman*. Every novel contains something which is unique to itself, any significant novel (and some much more than others) challenges the expectations and assumptions of readers to a certain extent, and many novels are intended by their authors to challenge and frustrate readers' expectations to some degree. I have already argued that the novel is a genre that is character-ized by flexibility and change – something that is connected to its association with the everyday, familiar world of its readers. As this world changes, so too will the novel in terms of both its form and its content.

The picaresque novel

disgrace

A 'picaro' in Spanish is a rogue, and the picaresque novel is built on the tradition of the sixteenth-century Spanish picaresque narrative, which typically portrayed a sharp-witted rogue living off his wits while travelling through a variety of usually low-life settings. Recent commentators have suggested that the key aspect of the picaro is that he is a minor delinquent who offends against moral and civil laws and whose behaviour is antisocial without being utterly vicious. The picaro typically lives by begging or by minor theft, he is cynical in his attitude to the softer emotions, especially love, and by witty and satirical comments questions established beliefs and customs. Critics frequently relate the emergence of the picaro to the break-up of the feudal world, and his open-minded, self-interested, geographical and intellectual wandering and questing are seen by many to merge into the spirit of possessive individualism that is ushered in with the growth of capitalist social relations.

The picaresque novel is typically *episodic*, lacking a sophisticated plot, and psychologically complex – or developing – characters.

Daniel Defoe's novel *Moll Flanders* is by no means a pure picaresque novel; quite apart from anything else the fact that its main character is a woman represents a significant divergence from previous picaresque novels. But it is possessed of many typical characteristics of the picaresque novel. It is constructed around a sequence of short 'episodes' none of which is longer than two or three pages, and it has little real development of character. (Moll repents when she is facing execution, but this seems based on no fundamental *change* in her character or values.)

Part of the appeal of the picaresque lies in its concern with the ordinary and the trivial, a concern which sets the picaresque apart from the romance, and which explains the historical importance accorded the picaresque in the history of the novel's emergence and development. For some commentators, however, the picaresque has a more than merely historical significance. Thus, although some argue that only a small number of sixteenth-century Spanish works are truly picaresque, a more catholic definition of the term allows us to say that there are picaresque elements in a range of both older and more recent novels.

The epistolary novel

An epistolary novel is told through letters ('epistles') exchanged between different characters. It flourished in particular in the eighteenth century. Samuel Richardson's *Pamela* (1740) and *Clarissa* (1747–8), Tobias

Smollett's *Humphry Clinker* (1771), and Fanny Burney's *Evelina* (1778) are classic epistolary novels. Writers in subsequent periods have made sporadic attempts to adapt the technique to their own needs, and one of the most successful examples of a recent epistolary novel is Alice Walker's *The Color Purple* (1982).

Given how important letters were in English society in the eighteenth century, when literacy was high among educated people but transport facilities were very primitive by modern standards, it is not surprising that the novel should have been heavily influenced by this form of communication in its early development. In his novel *Clarissa* Samuel Richardson gives a comment to his character Mr Belfield, writing to Lovelace about Clarissa, which helps to explain why Richardson (and his readers) favoured the epistolary technique:

> Such a sweetness of temper, so much patience and resignation, as she seems to be mistress of; yet writing of and in the midst of *present* distresses! How *much more* lively and affecting, for that reason, must her style be, her mind tortured by the pangs of uncertainty (the events then hidden in the womb of fate), *than* the dry narrative, unanimated style of a person relating difficulties and dangers surmounted; the relater perfectly at ease; and if himself unmoved by his own story, not likely greatly to affect the reader. (Italics in original)

Clearly the personal insight, self-display, and dramatic effectiveness of the epistolary technique are what appeal to Richardson. But a novel told exclusively through letters (Richardson allows himself the occasional authorial footnote and square-bracketed interpolation!) can be unwieldy and inflexible. The characters have to be kept apart (otherwise they have no reason to write to one another), and sometimes this involves artificiality: Richardson's heroines are banished to locked rooms with disturbing regularity. And when they actually meet their loved ones Richardson has somehow to arrange for someone else to write a letter about the encounter. Moreover, whatever else the characters do they must always have access to pen and paper; just as the subjects of Browning's dramatic monologues cannot be allowed to lose their voices, so too Richardson's heroines must keep writing. As Clarissa on one occasion writes to Miss Howe: 'And indeed, my dear, I know not how to *forbear* writing. I have now no other employment or diversion. And I must write on, although I were not to send it to anybody.' It was this element in Richardson's *Pamela* that was so cruelly but effectively satirized by Henry Fielding in his parody *Shamela*, in which the heroine continues to write a letter while the man who is assaulting her virtue climbs into bed with her.[1] On the other hand, the joke reminds us that one of the attractions of the technique is that the reader can feel that he or she has the very matter of the plot in his or her hands – that the experience of reading the novel duplicates the experiences which themselves constitute the novel.

If the 'pure' epistolary novel is rare after the eighteenth century, the technique taught novelists how very useful letters could be as an element within the narrative variety of a novel. Think how useful it is to Emily Brontë to be able to include Isabella's long letter about her marriage to Heathcliff in *Wuthering Heights* (1847), and how perfectly Helen's letters to her sister contribute to E.M. Forster's *Howards End* (1910). But perhaps the most important thing that later novelists were able to learn from the epistolary novel was that it is unwise to straitjacket a novelist by the choice of too restrictive or inflexible a narrative technique, and, as we shall see later on, the post-eighteenth-century novel generally displays a far greater variety and flexibility of narrative technique than is to be found in the classic epistolary novel.

The three-decker

During the Victorian period, publication of a novel in book form (rather than in instalments in a periodical) commonly involved the production of a three-decker, that is to say, a novel produced in three volumes. Such three-deckers were generally borrowed from one of the circulating libraries such as Mudie's. Publication aimed at personal sales usually followed three-decker publication, and was normally in the form of a single volume.

Such publishing conventions imposed constraints upon authors, who were expected to conform to them. Thus if a novel was first published in the form of a three-decker, or in instalments in a periodical, this is worth knowing, for it may explain some of the work's characteristics which otherwise might be hard to understand – or even to notice. It is useful to be able to read Henry James's *The Turn of the Screw* (1898) in an edition that marks the division between the instalments of its first published version, as we can then note that James attempts to end each separate instalment at a point of increased tension or dramatic surprise (much like

---

[1] The full title of Fielding's parody is: *An Apology for the Life of Mrs Shamela Andrews* (1741).

the old film serials for children, each episode of which typically ended with a 'cliff-hanging' finale).

The historical novel   *DISGRACE*

As its name suggests the historical novel sets its events and characters in a well-defined historical context, and it may include both fictional and real characters. It is often distinguished (in its more respectable forms) by convincing detailed description of the manners, building, institutions and scenery of its chosen setting, and generally attempts to convey a sense of historical verisimilitude. Its most respected practitioner in Britain has been Sir Walter Scott.

In its more recent popular, 'pulp' form it tends to abandon verisimilitude for fantasy, and in some ways can be seen as the present-day version of the romance, encouraging more an escape from contemporary reality than a critical and imaginative scrutiny of the life that its readers would recognize as their own.

The regional novel

The regional novel involves an especial focus of attention on to the life of a particular, well-defined geographical region. Traditionally the region in question will be rural rather than urban (thus it would be distinctly odd to refer to Charles Dickens as a regional novelist, in spite of the fact that so many of his novels explore London life in such intimate detail). Very often a 'regional novelist' will write a number of books all involving the same territory or place – as with Thomas Hardy's 'Wessex' and William Faulkner's 'Yoknapatawpha County'. Both of these 'regions' are closely modelled on particular areas of England and the United States, in spite of their fictitious names (and, ironically, the popularity of Hardy's novels has led to the term 'Wessex' being used to denote that part of England upon which it is based).

Whether we feel happy about calling a novelist a regional novelist may well depend upon our view of his or her stature: to many the term has a rather limiting ring, and this leads them to reserve use of it for indisputably minor novelists.

The satirical novel

Satire does not have to be either in prose or fictional, although there is a sense in which the exaggeration upon which it typically depends necessarily involves a certain amount of fictive imagining. There is a tradition of satire which is independent of the novel and which stretches back to antiquity, and aspects of this tradition have been important influences upon the novel. Satire attacks alleged vices and stupidities – either of individuals or of whole communities or groups – and its tools are those of ridicule, exaggeration, and contempt.

Jonathan Swift's satirical writings – and especially his *Gulliver's Travels* (1726) – are extremely important staging-posts for the novel, although most commentators have been unwilling to allow that any themselves merit the title novels. If we think of Mark Twain's *The Adventures of Huckleberry Finn* (1884) we can see how the mature novel of the nineteenth century was able to incorporate important satirical techniques from earlier writers within a much more realistic framework than Swift, for example, provides. The use of a naïve narrator travelling among people whose strange ways he describes innocently while at the same time conveying his creator's biting satire of what they stand for is one element which unites Swift's and Twain's major works.

The satirist is by definition more concerned to draw our attention to what he or she is attacking than to create characters, situations and events that are believable in and for themselves. A novelist may however include satirical elements in works that do not, overall, merit the term 'satirical novel' (and indeed most novelists do). Thus E.M. Forster's *Howards End* is not usually categorized as a satirical novel, but it does include a distinctively satirical vein in its treatment of various characters. Take the following passage from the end of the first paragraph of Chapter Three: Mrs Munt is considering her responsibilities towards the two Schlegel sisters, Helen and Margaret:

> Sooner or later the girls would enter on the process known as throwing themselves away, and if they had delayed hitherto it was only that they might throw themselves more vehemently in the future. They saw too many people at Wickham Place – unshaven musicians, an actress even, German cousins (one knows what foreigners are), acquaintances picked up at continental hotels (one knows what they are too). It was interesting, and down at Swanage no one appreciated culture more than Mrs Munt; but it was dangerous, and disaster was bound to come.

On one level this passage is a good example of *irony*: the surface meanings of the statements it contains are diametrically opposed to the underlying meanings which we attribute to narrator or author (Forster clearly does *not* believe that no-one in Swanage appreciates culture more than Mrs Munt!). But the passage is complicated by the fact that it is also satirical: it attempts to diminish a set of beliefs by making them appear ridiculous. Although the passage is ostensibly related to us from the viewpoint of Forster's narrator it mimics Mrs Munt and makes her views seem absurd by pretending to adopt them (Forster makes use of Free Indirect Discourse here – see p. 109.)

If we compare *Howards End* with Joseph Heller's *Catch 22* (1961) we will appreciate, I think, that the latter can more properly be termed a satirical novel than the former. Heller's dominating concern is clearly to diminish war and the military in our eyes by making them appear ridiculous and vain; his major aim in the novel is, in other words, satirical. *Howards End*, in contrast, contains satirical elements but it is not dominated by these – indeed, as the novel progresses they loom less and less large.

The *bildungsroman* (novel of formation or education)

The German term *bildungsroman* is now generally used in English to denote that sort of novel which concentrates upon one character's development from early youth to some sort of maturity. Goethe's *Wilhelm Meister's Apprenticeship* (1795–6), Dickens's *David Copperfield* (1849–50), and James Joyce's *Portrait of the Artist as a Young Man* (1916) can all be described as *bildungsroman*.

This type of novel clearly attracts the writer interested in depicting the close relationship between early influences and later character development, and its emergence can be related to the growing interest in the theme that 'the child is father to the man' that accompanies the late eighteenth- and nineteenth-century interest in the young.

The *roman à clef* ('novel with a key')

The *roman à clef* is the sort of novel that can be 'unlocked' given the right 'key' – in other words one which refers to real people, places or events in disguised form so that once one realizes what the work is about the hidden references all become apparent. Thomas Love Peacock wrote several humorous novels in the early nineteenth century in which thinly disguised portrayals of contemporary individuals such as Shelley and Coleridge appeared.

As with many of the terms under consideration here, a central question is whether a work is dominated by a particular characteristic or whether the latter plays only a minor or subservient function within it.

Thus we do not call a novel a *roman à clef* simply because it contains one thinly disguised (or even thickly disguised) portrayal of a real person. *Sons and Lovers* (1913) is not a *roman à clef* because Paul Morel has a lot in common with the real D.H. Lawrence. Thomas Mann's *The Magic Mountain* (1924) was much admired by the Hungarian Marxist critic Georg Lukács, who apparently did not recognize what is allegedly the case – that the highly unpleasant character Naphta in the novel was based upon Lukács himself. Again, it would be absurd to use this as a reason for calling *The Magic Mountain* a *roman à clef*.

The *roman à these / tendenzroman* (thesis novel)

The *roman à these* has, as the term suggests, a particular thesis or argument underlying it. It is typically a novel concerned to encourage social reform, or the correction of a particular abuse or wrong. Central to the definition is the idea of a *dominant* and usually a simple and uncomplicated thesis. (See p. 137 below for the distinction between 'thesis' and 'theme'.) Harriet Beecher Stowe's *Uncle Tom's Cabin* (1852), which is structured around an attack upon the institution of slavery in the United States, is a classic *roman à these*. A novel like Thomas Hardy's *Jude The Obscure* (1896) in contrast, although it undeniably involves an attack upon certain social conventions which are seen as repressive and although in a sense it clearly 'calls for' a change in society so that women and working people have greater opportunities for self-development and for education, is too complex and involved a novel to be termed a *roman à these*. A good test is to ask whether the aim of a novel can be summed up simply as 'An attack upon "X"'; if not, then it is unwise to refer to it as a *roman à these*. As these comments suggest, the term has a pejorative ring to many ears.

The *roman noir* / gothic novel

The more usual term in English is 'gothic novel'. The term denotes a type of fiction that was ushered in by Horace Walpole's *The Castle of Otranto* (1764). Walpole was much influenced by the revival of interest in the gothic that occurred in the later eighteenth century, a revival that can be seen as a precursor of aspects of Romanticism in its predilection for the wild, the uncanny, and the horrific – all of which were associated by the pre-Romantics with the medieval period. The gothic novel introduced stock characters, situations and settings that still survive in the modern horror film: gloomy medieval settings, ancient castles with secret rooms and passages ruled over by a sinister nobleman tortured by a guilty secret, and a strongly supernatural element. The gothic novel proper flourished for a limited period, and is more or less a thing of the past by the early nineteenth century, but gothic elements can be found in a wide range of fiction during the nineteenth and even twentieth centuries. Jane Austen parodies aspects of the gothic novel in *Northanger Abbey* (begun in 1798 and published in 1816). The term 'gothic' has been applied to the works of Mervyn Peake much more recently, and the survival of the gothic in popular film and literature probably guarantees that it will remain a continuing influence on serious literature.

The work of certain recent women writers such as Angela Carter and Sylvia Plath has been claimed to contain gothic elements, and a revival of interest in the gothic has been encouraged by a feminist interest in the way in which gothic formulae can be seen both to encapsulate certain stereotypical masculine fears about women and also to allow women the space to explore hidden aspects of gender formation.

The *roman-fleuve*

This term denotes a series of sequence of novels which can be read and appreciated individually but which deal with recurring characters and/or common events and which form a sequence or which complement one another. Perhaps the best-known example is Balzac's *La Comedie Humaine*, but Anthony Powell's *A Dance to the Music of Time* (1951–75) is a more recent example. The *roman-fleuve* is closely related to what is called the *saga novel* – a series of novels about a large family each of which concentrates upon different branches of the family or different events in which it is implicated. Thomas Mann's tetralogy *Joseph and his Brothers* (1933–43) is one twentieth-century example; John Galsworthy's *Forsyte Saga* (1906–35) is another.

The *roman feuilleton*

This is a novel that is published in instalments in unabridged form by a daily newspaper. The method of publication is unusual today for a novel, although it is becoming more and more fashionable for certain sorts of non-fictional work to be published in abridged form in newspaper instalments. In the nineteenth century it was not unusual for novels to be published in unabridged form in newspaper instalments.

Science fiction

Some definitions link science fiction with *fantastic* literature, and the two are clearly closely related. But whereas the fantastic normally invokes the strong possibility (but not certainty) of supernatural agencies, science fiction does not generally do this. Science fiction is rather characterized by settings involving interplanetary travel, advanced technology, and is typically set in the future. In contrast to fantasy literature its settings and events are often conceivable but not actual.

Jules Verne and H.G. Wells are often granted the joint title of the father of science fiction (its most celebrated authors tend to be male), and its best-known practitioners today are Ray Bradbury, Arthur C. Clarke, and Isaac Asimov.

It is possible to relate science fiction to a work such as Swift's *Gulliver's Travels*, and clearly science fiction can have a 'pastoral' element: commenting upon one society or community under the pretext of describing another, imagined one.

The postmodernist novel

During the last few years it has become clear that 'postmodernism' has emerged as the umbrella term which is most frequently used to describe a range of formal and technical experiments which are linked to theories about reality in general and the modern world in particular. In his *Postmodernist Fiction*, Brian McHale distinguishes postmodernism from modernism by differentiating between a poetics (or systematized theory of literature) dominated by ontological issues (that is, discussion of the nature of reality in general), and a poetics dominated by epistemological issues (that is, how and if we know the world or 'the real').[2] Thus the six entries which follow the present one in this chapter can all be seen as examples of, or variants upon, postmodernism. Postmodernism is by many seen to involve (among other things) the taking of certain modernist techniques to an extreme, and the discussion of modernism and postmodernism in Chapter 5 below should be consulted for further information. Prominent American postmodernists are Donald Barthelme, Ronald Sukenick, John Barth and Walter Abish, but British novelists such as B.S. Johnson and Christine Brooke-Rose, the Colombian Gabriel

---

[2] Brian McHale, *Postmodernist Fiction* (London, Routledge, repr. 1989, first publ. 1987), p. xii.

García Márquez, and many others are now regularly described as postmodernist novelists.

The *nouveau roman* (new novel)

In the *nouveau roman* the accepted conventions of fictional composition are deliberately distorted or flouted in order to disorient the reader and to foreground certain processes of artistic illusion or narrative convention. As such it can be seen as a post-Second World War development of *modernism*. It is associated particularly with France.

The best-known exponents of the *nouveau roman* are Alain Robbe-Grillet (whose book *Pour un Nouveau Roman* [1963] originated the term), Michel Butor, and Nathalie Sarraute.

Metafiction

Metafiction is, literally, fiction about fiction – normally denoting the sort of novel or short story which deliberately breaks fictive illusions and comments directly upon its own fictive nature or process of composition. The English father-figure of metafiction is Lawrence Sterne, in whose *Tristram Shandy* (1760–67) the narrator jokes with and teases the reader in various ways – advising him or her to turn back several pages to read a passage afresh, for example. Metafictional techniques are common to much modernist and, especially, postmodernist writing.

One of the best-known recent examples of such writing is John Fowles's *The French Lieutenant's Woman* (1969).

The fantastic

Characteristic of the fantastic is that it *hints* at a supernatural explanation without actually confirming that the supernatural has been involved in the events and experiences described. It thus constitutes one of the more recent movements away from a view associating the novel with a rather more traditional realism. According to some theorists – especially those influenced by Tzvetan Todorov – to be classified as a genuine example of the fantastic, a novel or short story must retain a sense of ambiguity to the end, never allowing the reader finally to establish that the depicted events can only be accounted for by natural or, alternatively, by supernatural, explanations. From this perspective there are few pure examples of the fantastic; according to some (but not all) accounts one would be Henry James's *The Turn of the Screw*. Recent interest in the fantastic has been associated with feminist writers and critics interested in exploring alternative 'realities' in which the gender conventions of past and existing societies are challenged or held up to scrutiny.

Fabulation and surfiction

These two terms are used relatively interchangeably. Both imply an aggressive and playful luxuriation in the non-representational, in which the writer takes delight in the artifice of writing rather than attempting to describe or make contact with a perceived extra-fictional reality. Both terms are normally used in connection with relatively recent fiction.

Magic realism

A blending of the fantastic or fantasy with realism, which today is particularly associated with certain Central and South American novelists such as Gabriel García Márquez and Isabella Allende but which has also been applied to some of the later work of the Kenyan novelist Ngũgĩ Wa Thiong'o and to various European novelists such as the German Günther Grass and the British Angela Carter – and even retrospectively to a novel such as Virginia Woolf's *Orlando* (1928) and to the female gothic novel. In Europe the term is particularly associated with certain recent female and feminist novelists. 'Blending' is perhaps misleading; magic realism seems typically to involve the sudden incursion of fantastic elements into an otherwise realistic plot and setting.

Faction

The term comes from the American author Truman Capote and is a portmanteau word (= fact + fiction) which describes novels such as his own *In Cold Blood* (1966). In this work primarily novelistic techniques are used to bring actual historical events to life for the reader. The term has thus come to denote a work that is on the borderline between fact and fiction, concerned primarily with a real event or persons, but using imagined detail to increase readability and verisimilitude.

The postcolonial novel

For postcolonialism in general, see the discussion on p. 175 in Chapter 8. As this account suggests, the term is both an ambiguous and a controversial one. Used to describe a particular sort of literature or fiction rather than a political position or a more general socio-historical period or geographical-political region the term is less problematic, and 'the postcolonial novel' is generally reserved to describe works of fiction written by citizens of lands which have recently thrown off colonial rule. It is normally the case that the description implies that the work(s) in question will take the immediate history of the land in question as a major part of its subject matter. Works emanating from much older ex-colonies such as Australia or the United States, or written about colonial lands by Europeans or North Americans, are less usually described as postcolonial.

4 The Short Story and the Novella

I have already suggested that our familiar tripartite division of prose fiction into the novel, the short story and the novella is one that simplifies a more complex and varied reality, and it is as well to start this chapter with a warning against allowing a use of these categories (more popular perhaps with academics than with writers and the common reader) to obscure the variety of forms open to the writer of prose fiction – both today and in the past. Having made that point, however, it is clear that the general acceptance which this tripartite division has won is indicative of a certain utility and convenience. The truth is that arguments about whether a particular piece of prose fiction is a novel or a novella – or a novella or a short story – are very rare. When they occur it is often as a result of the comparatively late acceptance of the novella as a separate generic category in the English-speaking world. For the rest of this chapter I will, then, continue to use these three categories while attempting to avoid imposing them on a more diverse reality in too mechanical a way.

If as I have claimed the novel is very much the child of print, one might assume that at least superficial resemblances between the short story and oral narratives – tales, anecdotes, and so on – might suggest that the short story has much older parents – which is in part what Mulk Raj Anand implies (see p. 46). But Frank O'Connor, one of the first major theorists of the short story, resists this idea. He argues that the short story, like the novel, 'is a modern art form; that is to say, it represents, better than poetry or drama, our own attitude to life'. This is not to say that he believes the short story to be merely a truncated novel: 'even from its beginnings, the short story has functioned in a quite different way from the novel'.[1] In his account of the nature of this difference, O'Connor is thought-provoking if, on occasions, eccentric. He suggests that, unlike

[1] Frank O'Connor, *The Lonely Voice*, p. 13.

the novel, the short story does not encourage identification between the reader and a character:

> In discussions of the modern novel we have come to talk of it as the novel without a hero. In fact, the short story has never had a hero. What it has instead is a submerged population group – a bad phrase which I have had to use for want of a better.
>
> Always in the short story there is this sense of outlawed figures wandering about the fringes of society, superimposed sometimes on symbolic figures whom they caricature and echo – Christ, Socrates, Moses. [. . .] As a result there is in the short story at its most characteristic something we do not often find in the novel – an intense awareness of human loneliness. Indeed, it might be truer to say that while we often read a familiar novel again for companionship, we approach the short story in a very different mood.[2]

These are large generalizations and we are right to be suspicious of large generalizations about categories of literary work which, as I have suggested, contain so much variety within themselves. Nevertheless they are thought-provoking generalizations. Is it not the case that the reader does often approach a short story and a novel with different expectations? And does this not have something to do with a lesser degree of identification with one or more of the characters on the part of the reader? Is the short story perhaps a genre that offers a more detached view of characters and events than do many novels?

One reason for this may be that, as another Irish critic of the short story, Sean O'Faolain, has argued, there can be no development of character in the short story (a point of which Henry James was, O'Faolain argues, unaware). Of course, characters in a short story may undergo shocking experiences and they may experience dramatic illuminations, such that they are not the same at the end as at the start of the story – and indeed many critics have suggested that such experiences are an almost essential component of the short story. But this is not at all the same as that gradual development of character that we can observe in a full-length novel.[3] For O'Faolain, at the heart of the short story has to reside the writer's ability to make fragments of experience speak for the whole of

[2] Frank O'Connor, *The Lonely Voice*, pp. 17–18, 19.

[3] Sean O'Faolain, *The Short Story* (Cork, The Mercier Press, 1972; first publ. London, 1948), p. 221.

life.[4] And this carries with it a range of technical necessities. In particular, the short story writer must be adept at replacing *direct telling* by *suggesting*:

> Telling by means of suggestion or implication is one of the most important of all the modern short-story's shorthand conventions. It means that a short-story writer does not directly tell us things so much as let us guess or know them by implying them. The technical advantage is obvious. It takes a long time to tell anything directly and explicitly, it is a rather heavy-handed way of conveying information, and it does not arrest our imagination or hold our attention so firmly as when we get a subtle hint. Telling never dilates the mind with suggestion as implication does.[5]

To make his point, O'Faolain presents his readers with a splendid analysis of the opening sentence of Anton Chekhov's short story 'The Lady with the Little Dog'. (It should be noted that O'Faolain believes that the openings of short stories in particular demand of the author that he or she be adept at influencing the reader by means of suggestion and implication.) The sentence in question reads as follows: 'It was reported that a new face had been seen on the quay; a lady with a little dog.' Here is O'Faolain's commentary:

> The amount of information conveyed in that sentence is an interesting example of the shorthand of the modern short-story. What do we gather from it? [. . .] We gather, altogether by implication, that the scene is laid in a port. We gather that this port is a seaside resort, for ladies with little dogs do not perambulate on commercial docks. We gather that the season is fine weather – probably summer or autumn. We gather that this seaside resort is a sleepy, unfrequented little place: for one does not observe new faces at big, crowded places like Brighton or Deauville. Furthermore, the phrase 'it was reported' implies that gossip circulates in a friendly way at this sleepy resort. We gather still more. We gather that somebody has been bored and wakes up at this bit of gossip; and that we shall presently hear about him. I say 'him', because one again guesses, when it is a question of a lady, that the person most likely to be interested is a man.[6]

---

[4] Sean O'Faolain, *The Short Story*, p. 185.

[5] Sean O'Faolain, *The Short Story*, p. 177.

[6] Sean O'Faolain, *The Short Story*, pp. 177–8.

As he adds (perhaps redundantly, in view of the contrast between the length of his commentary and the length of the sentence in question!), 'we may imagine how much time it would take, and how boring it would be, to have all that told at length'.

When I first read this passage in O'Faolain's study I immediately started pulling novels off the shelf to try to disprove his point by finding a novel that opened with a sentence as full of suggestion as is Chekhov's. After over ten tries, the best I could find was the opening to Alan Sillitoe's *Saturday Night and Sunday Morning* (1958): 'The rowdy gang of singers who sat at the scattered tables saw Arthur walk unsteadily to the head of the stairs, and though they must all have known that he was dead drunk, and seen the danger he would soon be in, no one attempted to talk to him and lead him back to his seat.' This imparts quite a lot, but it is longer than the first sentence of 'The Lady with the Little Dog', and less compressed and economical. Extending the survey to a range of short stories selected at random convinced me that, although exceptions could be found, O'Faolain was probably right in the broad distinction he draws between the novel and the short story.

Now we should recognize, at this point, that to a certain extent all literature works by implication and suggestion. And it would be possible to give a long list of comments from various novelists about precisely this – how prose fiction works on the reader by stimulating his or her imagination through suggestion or implication. Yet, even so, O'Faolain may well be right that the writer of short stories has to be even more adept at planting small seeds that will grow into large oaks in the minds of readers than does the novelist. A novelist can repeat a point if it is felt that the reader may have missed it: the short story writer, however, must get it across the first time: there is no space for a second attempt. The reduced length of the short story, in other words, carries with it certain formal and technical requirements.

The short story is normally read at one sitting – Edgar Allan Poe in fact suggested that this was a necessary feature of the short story. Because of this the reading of a short story tends to be less reflective and more concentrated an experience; we rarely stop several times for prolonged thought in the middle of reading a short story. The plot of a short story typically limits itself to a brief span of time, and, as O'Faolain argues, rather than showing its characters developing and maturing will show them at some revealing moment of crisis – whether internal or external. Short stories rarely have complex plots; again the focus is upon a particular episode or situation rather than a chain of events.

Thus much of the skill of the short story writer has to be devoted to making characters appear three-dimensional in spite of the fact that we see them for only a very short period of time. In addition care has to be taken to render atmosphere and situation convincingly. Very often the short story writer will use something akin to shock tactics to make the reader think and respond: an unexpected ending, a dramatic unveiling, a surprising twist of plot. Many consider these to be artistically inferior techniques, less sophisticated than other means of creating depth and body in the short story. What are these means?

We have already mentioned *suggestion*, and clearly there are a number of techniques whereby we get people's imaginations racing. At the risk of over-simplifying a complex issue we should probably remember that there are two sorts of suggestion: the first when we know exactly what it is that we want the target person to think (as Iago gets Othello to imagine that Desdemona has been unfaithful without actually stating as much), and the second when a person's imagination is stimulated to be innovative and to think of things that the suggester him- or herself has never intended. There seems little doubt that the second of these alternatives is potentially richer and more aesthetically rewarding, as it gives the reader's imagination a genuinely *creative* function.[7] For me a classic example of such suggestiveness used to full aesthetic effect is to be found at the end of James Joyce's 'The Dead' (started 1907, published 1915), one of the stories from his *Dubliners*. At this point of the story the character Gabriel, in a hotel room with his sleeping wife after the annual dance held by Gabriel's aunts, is sinking into sleep. The final lines of the story read:

A few light taps against the pane made him turn to the window. It had begun to snow again. He watched sleepily the flakes, silver and dark, falling obliquely against the lamplight. The time had come for him to set out on his journey westward. Yes, the newspapers were right: snow was general all over Ireland. It was falling on every part of the dark central plain, on the treeless hills, falling softly upon the Bog of Allen and, farther westward, softly falling into the dark mutinous Shannon waves. It was falling, too, upon every part of the lonely churchyard on the hill where Michael Furey lay buried. It lay thickly drifted on the crooked crosses and headstones, on the spears of the little gate, on the barren thorns. His soul swooned slowly as he heard the snow falling faintly

[7] For a useful discussion of this topic, see Krishna Rayan, *Suggestion in Literature* (London, Edward Arnold, 1987).

through the universe and faintly falling, like the descent of their last end, upon all the living and the dead.

The suggestiveness of this very rich passage is partly a matter of inexplicitness: what for example does 'The time had come for him to set out on his journey westward' mean? We have earlier learned that Gabriel does not want to accompany his wife on a visit to the west of Ireland. Has he changed his mind, or is the reader supposed to recall that the idiom 'to go west' means to die, and to associate this with sleep as proleptic of death? This is my second sort of suggestiveness, one that is creative and open-ended rather than closed and with a single, fixed objective.

Those who have read 'The Dead' will recognize that in this final paragraph a shift to what we can call a different mode takes place; the language becomes more poetic, and the more literal meanings of certain statements are hard to determine, while at the same time there is uncertainty as to who is making them. The ending makes extremely difficult demands on the reader, and is suggestive in a number of different ways (to take a simple example, the mention of 'crosses', 'spears' and 'barren thorns' inevitably calls the Passion of Christ to mind, a very different passion from the one experienced by Michael Furey although one which also ended in death and which thus fixes a certain association between passion and death that has already been implied).

In addition to being suggestive the ending is also evocative. That is to say, it calls up particular states of mind and forms of experience, very often in their concrete particularity. Having said this, it is less easy to say exactly what it is that is evoked. There is a general movement in the paragraph towards a greater and greater universality, fixed in the actual mention of the word 'universe' in the last sentence. It is as if we leave Gabriel (whose consciousness we depart from as the vision becomes more and more general), and take more and more in to our view. But as this shift coincides with Gabriel's falling asleep, one of the experiences evoked is precisely that of falling asleep, leaving more rational, specific topics of thought and entering into a dream-state that deals with more general and universal topics. Note that what I have written about this passage is necessarily tentative. Necessarily, because if one is dealing with suggestion then certainty is no longer possible.

The last paragraph of 'The Dead' illustrates other important techniques. The use of poetic devices such as repetition and chiasmus (reversal involving the pattern 'ab ba', as in the opposition of 'falling faintly' and 'faintly falling'), symbolism (see the discussion of symbol and image in Chapter 6, which pays particular attention to the repeated references to snow in 'The Dead'), and other techniques are not limited to the short story, but they very often have to carry a heavier weight in this genre to compensate for the limitations that a restriction of length places on the development of plot and character.

The *novella* has had less theoretical attention devoted to it than has the short story, and especially in its modern and its Anglo-Saxon manifestations. It is usually rather longer than a short story, and would not normally be read in one sitting. Joseph Conrad's *Heart of Darkness* can be seen as a classic novella, and any reader of this work will probably understand that the terms 'novel' and 'short story' seem somehow inappropriate to describe it.

Theories of the novella are often constructed with particular reference to the German novella tradition as the novella has enjoyed a greater prominence in Germany than it has elsewhere in Europe, and as a result one needs to guard against making large generalizations on the basis of a limited sample of novellas. However, it does seem to be the case that the novella often has a dominant symbol or complex of symbols at its heart, and that it is these rather than the complexity of its plot that give the novella its depth and significance. As with the *nouvelle* the novella is typically more limited in its concerns, often restricting itself to a single state of affairs, set of relationships, or setting. It thus has some of the concentrated power of the short story, but without the frequent one-dimensionality that characterizes many short stories. It is hard to imagine many short stories that could have generated the enormous body of interpretative responses that *Heart of Darkness* has done, and this has something to do with the symbolic richness of many novellas. I have said that generalizations are dangerous, however, and the attentive reader will spot that what I have just said can hardly be reconciled with my earlier comments about the suggestiveness of the short story and my comments on the ending of Joyce's 'The Dead'. I would venture, however, that 'The Dead' probably *is* unusual among short stories in terms of the richness and intensity of its symbolism, and it will also be noted that it is relatively long for a short story. I suspect that not too many eyebrows would be lifted were one to describe 'The Dead' as a novella: it is clearly nearer the border-line between the two forms than is Chekhov's 'The Lady with the Little Dog'. Furthermore, although a short story may be suggestive in O'Faolain's sense, and this may lead to a rich reading experience, it does not necessarily involve the same potentiality for interpretative complexity that we find in a major novella such as *Heart of Darkness*.

The actual events of *Heart of Darkness* could be summarized in a few lines; whatever we read this work for it is not for complexity of plot development.[8] Instead we need to pay attention to Conrad's use of symbol and image in the work, to the complexities of narrative technique (the 'tale within a tale' and the interplay between the frame and framed narrative, each of which has its own narrator), and to the texture of the prose. The novella is dominated by Marlow (narrator of the framed narrative), and his fascination with, and search for, the mysterious figure of Kurtz. Early on in the story Marlow comes across a painting that Kurtz has executed:

> Then I noticed a small sketch in oils, on a panel, representing a woman, draped and blindfolded, carrying a lighted torch. The background was sombre – almost black. The movement of the woman was stately, and the effect of the torch-light on the face was sinister.

We see here, I think, how a relatively short work can be given a complex unity through the use of means other than those involving the intricacies of plot development. The picture is one of a number of symbolic 'moments' in the work which draw various threads together in a masterly way. We note a relationship between the blindfolded woman carrying a lighted torch and those Europeans (including Kurtz) who have claimed that they are bringing light to Africa but who are actually 'going at it blind', either self-deceived or dishonestly plundering. We are reminded of Kurtz's fiancée 'the Intended' – with whom we meet at the end of the tale and who is as blind metaphorically as the painted woman is literally. (As Marlow's account makes clear, European women have to be left in the dark about what their men are doing in Africa at this time – including what they are doing with African women such as Kurtz's mistress.) The references to light and blackness link up with a repetitive pattern of black-white images in the work which have an intricate relationship to sets of moral judgements indicated by Conrad in indirect ways.

Writers of short stories and novellas, in other words, do not just produce truncated novels. These different fictional genres require a different use of the resources of the writer of prose, and they should not be read or judged in the way that we read or judge a 500-page novel.

---

[8] Although in his *The Deceptive Text: An Introduction to Covert Plots* (Brighton, Harvester, 1984), Cedric Watts argues that *Heart of Darkness* has a complex plot structure consisting of both an overt and a covert plot.

5 Realism, Modernism, Postmodernism

In addition to the terms discussed in the previous chapter denoting particular types of novel there are three terms which refer to something wider than a type of novel or short story that it will be as well for us to consider at this point: *realism*, *modernism*, and *postmodernism*. These terms are used not just in connection with the discussion of literature or of fiction but have an application to more widespread tendencies within the culture of (primarily) the past two centuries.

The first two of these terms are doubly complicated by the fact that each can be used both to refer to *historical periods of literature* and also to transhistorical *types of literature* (although the term 'modernism' is not normally applied to literature written before the final years of the nineteenth century). In this respect the problems associated with the use of these terms are not dissimilar to those attached to the use of the term 'romanticism', another term which can also be used to denote both a tightly delimited period of literature and also a particular variety of literature. The use of 'romanticism' too, is also complicated by the fact that it is not just to literature that it is used to refer.

Just as not all literature written during the romantic period is necessarily romantic, so too not all modern fiction is modernist, and indeed much fiction written today is neither modernist nor postmodernist. Moreover, just as there are arguments about what does or does not characterize romanticism, so too there is by no means any universal agreement as to what constitutes realism, modernism, or postmodernism in the novel.

Given all these complications it might be thought wise merely to avoid using any of these terms. This simple solution is not, alas, a viable one. The history of the novel is so intimately bound up with the issue of realism that we can hardly talk about the novel without addressing it.

Moreover, that there is a family of new characteristics to be found in much modern art and literature can scarcely be disputed, and the term 'modernism' allows us to isolate these characteristics and to distinguish between radically different lines of development in the art and literature of the present century. Finally, although theories of the postmodern have come into major prominence only during the last couple of decades, they are so central to discussion not just of recent experimental fiction but also of many much older works that they need to be considered.

We saw in our earlier discussion of the emergence and development of the novel that the genre is distinguished by what we can call its 'formal realism': it is stocked with people and places that *seem real* or *evoke the real* even if they are imagined, whereas genres such as the romance or the epic were peopled with people and places that seem (and were meant to seem) unreal and which do not turn our attention to everyday reality. We can note, furthermore, that 'realism' is a term which a large number of novelists have felt the need to use in connection with their work.

On a simple level it can be said that something – a character, an event, a setting – is 'realistic' if it resembles a model in everyday life. The matter becomes more complex, however, when we remember that a novelist can make us think seriously and critically about the real world by creating characters, events and settings that in many ways diverge from what we would expect in everyday life – in 'reality'. Peter Lamarque has noted that a 'fiction is realistic if it describes characters with combinations of properties that would not be strange or out of place if exemplified in individuals in the real world', but he goes on to discuss the problems raised by a novel such as George Orwell's *Animal Farm* (1945) which, in spite of the fact that it contains farmyard animals which speak, think and reason exactly like human beings, also strikes us as 'realistic' or 'true to life'. He suggests that what we need to explain this apparent contradiction 'is an appeal to levels of interpretation or understanding'.[1]

In other words, it is difficult to arrive at a concept of realism which is not trivial unless one goes beyond a concern with simple resemblance and proceeds to consider not just *what is in*, say, a novel but also *how a reader responds* to this content and its presentation.

We need to remember how many unusual coincidences and extraordinary events there are in those novels which are normally described as highly realistic: George Eliot's *Middlemarch* (1871/2) for example. But Lamarque's mention of *Animal Farm* should also draw our attention to

---

[1] Peter Lamarque, *Fictional Points of View*, p. 38.

the issues raised by parody and satire, both of which can be used by the novelist to make us think about the real world we inhabit. A parody is often very unrealistic in one sense, depending upon such extremes of exaggeration and such pointed selection that it cannot be said to give us anything that can be compared to our everyday world *in terms of direct, one-to-one resemblance*. But parodies and satires clearly *do* cause us to see our everyday world very differently. The paradox, then, is that sometimes that which distorts what the real world is like actually leads us to see the real world more accurately.

One of the most influential debates about the nature of realism in the twentieth century involved a critic very much associated with the novel – Georg Lukács. For Lukács, who saw realism as the contemporary artist's necessary, even ultimate, aim, the artist was required to portray the totality of reality at any given point in history, to penetrate between surface appearances, and to reveal processes of change. He believed that the artist's task was similar to that set for himself by Marx: to understand world history as a complex and dynamic totality through the uncovering of certain underlying laws. In practice this led Lukács to place a supreme value upon certain works of classical realism in the novel – the names of Tolstoy, Balzac and Thomas Mann are frequently on his lips or at the tip of his pen – and to wage an unceasing campaign against different aspects of what he classified as modernism (see below). At the end of his *The Historical Novel* (written 1936–7) Lukács referred to the 'misunderstanding' of his position that 'we intended a formal revival, an artistic imitation of the classical historical novel',[2] but even so many commentators have seen his opposition to modernism to be so all-embracing that this is actually what would have been required in order to satisfy his requirement that a literary work be realistic. Lukács's views are important for us to know about, because whatever his intention (and this is still a matter for debate), his arguments tended to encourage a view of the 'classic realist novel' of the nineteenth century as the pinnacle of artistic achievement of the novel, from which point on a decline took place as novelists abandoned that which the genre demanded of them. As this shift from realism to modernism coincides with the 'serious' novel's loss of its unambiguously popular status, the common reader has been dragged into discussions about the proper rôle and form for the novel.

---

[2] Georg Lukács, *The Historical Novel*, (trans.) Hannah and Stanley Mitchell (Harmondsworth, Penguin Books, repr. 1969), p. 422.

Thus when the German dramatist Bertolt Brecht, writing about the concept of realism, stated that we 'must not derive realism as such from particular existing works',[3] the unstated target was clearly Lukács (especially as he goes on to refer to Balzac and Tolstoy). Brecht's argument is, in a nutshell, an anti-essentialist one. In other words, realism for him is not intrinsic to a literary work, born with it like a genetic code and remaining unchanged through all its vicissitudes, but a function of the rôle the work plays or can play – the effect it has on a reader's view of the world. Brecht's realism focuses not on questions of form or content, but of function. The important point to bear in mind is that, when a critic discusses whether or not a novel is realistic, he or she can be referring either to what it contains or to what it does. Now very often critics and lecturers use the term in a loose sense and restrict it to the question of verisimilitude, but we should remember that a novel can be unrealistic in the former sense but realistic in the latter – or vice versa.

Think of it this way: if what we read in a novel seems just like the world we experience every day, it may be unlikely to encourage us to examine that world critically, to see it in a new light, to ask whether its appearance is deceptive. In contrast, if a novel presents us with a world that seems extremely unusual, it *may* encourage us to ask to what extent our familiar world is really like this, in ways we have not previously considered. Novels that are most like our everyday reality are not always the ones that lead us to think most critically about that reality.

One other, related point: most present-day commentators agree that James Joyce's *Ulysses* (1914–22) manages to get closer to how human beings actually experience the world than did the works of most of his novel-writing contemporaries. Yet many (but not all) of those who read this novel upon its first publication did not see it this way: they thought that it presented an idiosyncratic, even perverted set of consciousnesses. What can we learn from this? Perhaps that one of the achievements of major art is to get us to recognize ourselves and our experiences, rather than to see these in terms conditioned by unrecognized conventions. If this is so, then we must confront a paradox: very often that which is realistic will at first appear unrealistic. (We need also to note that, as

[3] Repr. in 'Brecht against Lukács', in Ernst Bloch, Georg Lukács, Bertolt Brecht, Walter Benjamin and Theodor Adorno, *Aesthetics and Politics* (London, New Left Books, 1977), p. 81.

Roman Jakobson has pointed out in a classic essay, art which appears realistic to one generation may appear artificial to a subsequent one.[4])

We have seen that the modern novel emerges in some sharp contrast to the romance, and is more realistic than the romance in the important sense that it directs its readers' attention towards the real world rather than offering an escape from the real world (and this is true, paradoxically, even when the actual reading of a novel can be a pleasant escape from immediate concerns and problems). But romance-like elements, as I have already suggested, can be used to explore our sense of the real: think of Swift's use of fantastic details in *Gulliver's Travels*, of James's suggestion of the supernatural in *The Turn of the Screw*, and of the use made by science fiction writers of imagined worlds.

The term 'realism' very often implies that the artist (and I repeat that this is not limited to the literary artist) has tried to include a wider and more representative coverage of social life in his or her work, and in particular that he or she has extended the coverage of the work to include 'low life' and the experiences of those deemed unworthy of artistic portrayal by other artists. The apparent realism of the early novel was intimately related in the public mind with the fact that it often concerned itself with the lives of the sort of human being who would not only never have been portrayed by the romance but whose depiction was clearly out-of-bounds too for the eighteenth-century poet. Fielding's Joseph Andrews, Defoe's Moll Flanders, Smollett's Humphry Clinker – none of these characters or their real-life equivalents could have entered into the polite world of Alexander Pope except in certain forms of satire or burlesque. And this is one of the reasons why *Joseph Andrews* (1742), *Moll Flanders*, and *Humphry Clinker* are generally taken to be more realistic works than 'The Rape of the Lock'.

'Realism' also has a specific reference to a particular literary movement which started in France in the early nineteenth century, and flourished in the latter part of the century. The names of the novelists most associated with this movement are those of Balzac, Stendhal, and Flaubert. These writers made enormous efforts to ensure that 'factual details' in their works were 'correct' – that is to say, capable of being checked against an external reality by empirical investigations. They achieved this accuracy by lengthy and painstaking research. 'Realism' in connection with these writers is thus both a term denoting a group of

---

[4] Roman Jakobson, 'On Realism in Art', (trans.) Karol Magassy. In Ladislav Matejka and Krystyna Pomorska (eds), *Readings in Russian Poetics: Formalist and Structuralist Views* (Cambridge, Mass., MIT Press), pp. 38–46.

novelists and also a term referring to a particular *method* of composition. (See also the comments on 'naturalism' below.)

The British heirs of the French realists included Arnold Bennett and George Moore – the former attacked by Virginia Woolf in essays such as 'Mr Bennett and Mrs Brown' for being a 'materialist' who was more interested in external details than in that inner life which, according to her, 'escapes' from his works.

One conclusion that can be drawn from the foregoing is that the term 'realistic' should be used only with extreme care in connection with a novel; it is a problematic rather than a self-obvious term and raises complex questions about what 'reality' is (our fantasies are after all real in the sense that they really exist and are related to our experiences in the real world), and about the means whereby a novelist explores the real. But having said all this it remains important to be able to distinguish between novels that – in however complex and indirect a way – cause us to think about reality critically and those which encourage us not to try but rather to escape from reality into a world of illusory imaginings.

Naturalism is a variety of realism, and the term is used in much the same way but with a narrower focus. Strictly speaking, naturalism should be restricted to a description of those literary works which were written according to a method founded upon the belief that there is a natural (rather than supernatural or spiritual) explanation for everything that exists or occurs. Naturalist novelists include the Goncourt brothers, Émile Zola, George Moore, the German Gerhard Hauptmann, and Americans such as Theodore Dreiser and Stephen Crane.

One more, important point needs to be reiterated at this juncture. It will be seen that Brecht's views, to which I referred above, open the way to the belief that a particular set of novelistic conventions can be productive of a realistic work at one point of time, but, because they become familiar and present the reader with no new perspective on the world, can lose that power at a later stage. Thus we should not, today, make the mistake of dismissing naturalist (or other) theories and methods as unsound merely because in our own time we believe them to be insufficient as a basis for producing fiction that will encourage readers to see the world in a new manner. We should try, rather, to find out what their force was at the time that they were formulated, and remember that the way we see the world may owe something to the effect that they had on readers in the past.

It should perhaps be added that, although twentieth-century fiction is often divided in very broad terms into 'realist', 'modernist' and 'post-modernist' fiction, this should not be taken to imply that fiction that is neither modernist nor postmodernist is somehow old-fashioned and un-modern, although it is true that many writers and critics have assumed and argued that this is the case. Moreover, the survival of realist conventions (a plot based upon cause-and-effect, well-defined characters, a general assumption that the world is knowable and susceptible to rational enquiry) in much popular fiction gives food for thought. One can posit either that the common reader is old-fashioned in his or her tastes and wishes to be presented with a world that is seen in ways that are familiar and thus undemanding, or alternatively that those writers and readers who produce and consume modernist fiction have another view of reality – a different world-view – from the mass of people who read detective stories, science fiction, and popular romance.

Modernism is a term which has come into more general use only since the Second World War, although (as with the term 'postmodernism') its history predates its general acceptance by many years. It refers to those art works (or the principles behind their creation) produced since the end of the nineteenth century which decisively reject the artistic conventions of the previous age. Foremost among such rejected conventions are those associated with realism in its more straightforward sense. In particular, modernist works tend to be *self-conscious* in ways that vary according to the genre or art-form in question; they deliberately remind the reader or observer that they *are* art-works, rather than seeking to serve as 'windows on reality'. Whereas one may forget that one is reading a novel when immersed in, say, *War and Peace* (1865–68) – responding to characters and events as if they were 'real' – this is something of which one is constantly reminded when reading a modernist work such as Virginia Woolf's *The Waves* (1931). Thus Picasso's rejection of representational art and of the conventions of perspective in his early paintings can be compared with the rejection of the 'tyranny of plot' by novelists such as Joyce, Woolf, and the Frenchman Marcel Proust. A good example of this shift in attitudes from realism to modernism can be found in the novels and other writings of Joseph Conrad, who in many ways represents a transitional stage between realist and modernist conventions. In 1898 Conrad wrote to his friend Cunninghame Graham, 'You must have a *plot*! If you haven't, every fool reviewer will kick you because there can't be literature without plot.'[5] And four years later, in 1902, he wrote to Arnold Bennett (one of Woolf's 'materialists' and thus far from what we now

[5] Frederick R. Karl and Laurence Davies (eds), *The Collected Letters of Joseph Conrad* (Cambridge, Cambridge University Press, 1986), volume 2 (1898–1902), p. 5.

call modernism), 'You just stop short of being absolutely real because you are faithful to your dogmas of realism. Now realism in art will never approach reality.'[6]

What we see here is 'the tide on the turn'; a writer beginning to question 'dogmas of realism' and to search for alternatives: alternatives to the well-made plot, the rounded and lifelike character, the knowable world wholly accessible to reasoned and rational enquiry.

The modernist novel typically focuses far greater attention on to the states and processes *inside* the consciousness of the main character(s) than on to public events in the outside world. If the twentieth century is the century of Freud and Marx then we can say that the modernist novel has much more in common with the former than the latter (which explains, incidentally, why an orthodox Marxist critic such as Georg Lukács argued so bitterly for realism and against modernism during his long life). Moreover, as I will go on to argue in greater detail in Chapter 6, modernism has a profound effect on novelists' conception of *character*. Michael Levenson opens his book *Modernism and the Fate of Individuality* as follows:

> This thing we name the individual, this piece of matter, this length of memory, this bearer of a proper name, this block in space, this whisper in time, this self-delighting, self-condemning oddity – what is it? Ours may be the age of narcissism, but it is also the century in which ego suffered unprecedented attacks upon its great pretensions, to be self-transparent and self-authorized.[7]

For the great modernist novelists (Levenson pays particular attention to Conrad, James, Forster, Ford Madox Ford, Wyndham Lewis, Lawrence, and Woolf) character can no longer be taken to be self-transparent – more than one character in a modernist novel asks 'who am I?' without receiving a clear answer. Nor is the self any longer its own source of authority – which rarely means that any alternative source of authority is readily available.

This focusing upon the problems of the self and of the inner life has encouraged the development of new *methods* of fictional expression. If

---

[6] Frederick R. Karl and Laurence Davies (eds), *The Collected Letters of Joseph Conrad*, volume 2 (1898–1902), p. 390.

[7] Michael Levenson, *Modernism and the Fate of Individuality: Character and Novelistic Form from Conrad to Woolf* (Cambridge, Cambridge University Press, 1991), p. xi.

modernism can be defined negatively in terms of its rejection of realist conventions and assumptions, its positive side can be seen in its remarkable development of techniques such as *stream of consciousness* and *internal monologue*, its challenging of traditional conceptions of story and plot, its markedly greater emphasis upon what Joyce calls 'epiphanies' and Virginia Woolf 'moments' – that is, points in time when reality seems to stand revealed and to speak itself – and its revolutionary use of various forms of what we can call 'poetic expression' in the novel. (Witness my earlier discussion of the final paragraph of Joyce's 'The Dead'.) Such techniques will be discussed in Chapter 6.

A brief comment should be added concerning the philosophical underpinnings of modernism. These are very often implicit rather than overt, but frequently we find that modernist novels are pessimistic in tone, unsure about the sense or logic of the world, and look on human beings as isolated and alienated. The philosophical corollary of the rejection of perspective in art, and of an omniscient view of a knowable world obeying certain laws in fiction, seems to be a view of reality as lacking any unifying logic, to the extent that one should perhaps talk rather of 'realities' than of reality. (Again risking a generalization, we can say that modernists have a monist view of reality but accept that complete knowledge of this reality is impossible, while postmodernists adhere to a pluralist view and deny that it makes any sense to talk of a single reality.) Thus, for the modernist, different perspectives have to be combined because although what they reveal may appear contradictory there is no 'super-perspective' by means of which to rank their validity.

With the term *postmodernism* we move on to other terminological problems. One perhaps inevitable confusion has arisen from the fact that, although one commentator has traced this term back to the mid-1930s,[8] it has only really come into more common use in Europe and the United States in the last decade or so. Thus works which before this time were categorized as either 'experimental' or 'modernist' have in some cases received a new, often disputed, categorization as postmodernist. As

[8] See Ihab Hassan, 'The Culture of Postmodernism', *Theory Culture and Society* 2(3), 1985, pp. 119–31. A letter from Charles Jencks to the *Times Literary Supplement*, 12 March 1993, points out that the term was used by the British artist John Watkins Chapman in the 1870s, and by Rudolf Pannwitz in 1917, and he also notes that the term 'modernism' was apparently coined in the third century!

Andreas Huyssen puts it, 'one critic's postmodernism is another critic's modernism'.[9]

Yet other commentators have refused to accept that the term is worth its salt, arguing that it does not really isolate any significant characteristics (in art, literature, or culture) that cannot be covered by the term 'modernism'. Such individuals frequently make what amounts to a political gesture by pointedly referring to 'late modernism' rather than 'postmodernism': behind the debate is the issue of whether we now inhabit a new sort of reality – the 'postmodern world' – which is fundamentally different from the world, or social systems, that produced the great modernist art. This political edge to the debate is worth noting, as the term *postmodernism* is typically used in a rather wider sense than is *modernism*, referring to a general human condition, or society at large, as much as to art or culture (a usage which was encouraged by Jean-François Lyotard's book *The Postmodern Condition: A Report on Knowledge* [English translation, 1984]). Not surprisingly, traditional Marxists, who believe we live within the same old capitalist system, are suspicious of the implications of the term 'postmodernism'.

Postmodernism, then, can be used today in a number of different ways: (i) to refer to the non-realist and non-traditional literature and art of the post-Second World War period; (ii) to refer to literature and art which takes certain modernist characteristics to an extreme stage; and (iii) to refer to aspects of a more general human condition in what is tendentiously referred to as the 'late capitalist' world of the post-1950s which have an all-embracing effect on life, culture, ideology and art, as well as (in some but not all usages) to a generally welcoming, celebrative attitude towards these aspects.

Those modernist characteristics which may produce postmodernism when taken to their most extreme forms include the rejection of representation in favour of self-reference – especially of a 'playful' and non-serious, non-constructive sort; the willing, even relieved, rejection of artistic aura (sense of 'holiness') and of a view of the work of art as organic whole; the substitution of confrontation and teasing of the reader for collaboration with him or her; the rejection of 'character' and 'plot' as meaningful or artistically defensible concepts or conventions; even the rejection of meaning itself along with the belief that it is worth trying to understand the world (or that there is a world to understand).

[9] Andreas Huyssen, *After the Great Divide: Modernism, Mass Culture and Postmodernism* (London, Macmillan, 1988), p. 59.

As Christopher Butler has suggested, many postmodern texts are baffling to those who (like critics!) wish to find coherence in, or impose it on, them by means of 'coherence-conferring strategies'. He continues:

> In reading a story like Donald Barthelme's 'The Indian Uprising', we cannot say 'who' the Indians are, what they 'symbolize', why the narrator makes tables out of hollow-cored doors while living with various women, why Jane is beaten up by a dwarf in a bar in Tenerife, as notified to the narrator by International Distress Coupon, and so on. And even if we could explain these elements separately in symbolic terms, there seems little chance that such explanations would be compatible with one another.[10]

Postmodernism takes the subjective idealism of modernism to the point of solipsism, but rejects the tragic and pessimistic elements in modernism in the apparent conclusion that if one cannot prevent Rome burning then one might as well enjoy the fiddling that is left open to one. A list of characteristics such as this has led some commentators to claim that certain much earlier works are postmodernist: the fiction of Franz Kafka, Knut Hamsun's *Hunger*, even Laurence Sterne's *Tristram Shandy*. I should however say that I am unhappy with such claims, as they obscure what I see as crucial differences between modernism and postmodernism. It is also possible to argue that there are postmodernist elements in the work of various post-structuralist and deconstructive critics such as Jacques Derrida, Michel Foucault, and Jacques Lacan (see Chapter 8).

Postmodernism is characterized in many accounts by a more welcoming, celebrative attitude towards the modern world. That this world is one of increasing fragmentation, of the dominance of commercial pressures, and of human powerlessness in the face of a blind technology, is not a point of dispute with modernism. But, whereas the major modernists reacted with horror or despair to their perception of these facts, in one view of the issue it is typical of postmodernism to react in a far more accepting manner. The extent to which postmodernism is mimetic of recent and new social, economic, and political practices in the societies in which it appears is a matter of some debate. Some have

---

[10] Christopher Butler, 'The Pleasures of the Experimental Text', in Jeremy Hawthorn (ed.), *Criticism and Critical Theory* (London, Edward Arnold, 1984), p. 131. For 'The Indian Uprising', see Donald Barthelme, *Unspeakable Practices, Unnatural Acts* (New York, Farrar, Straus & Giroux, 1976).

argued that the different but uncommunicating worlds with which many a postmodern novel faces the reader, reflect the fact that more and more people in the modern world live in self-contained ghettoes about which outsiders know little or nothing.[11] Others have suggested that perhaps they are reflective only of the contemporary isolation of certain artists and intellectuals, who then generalize their experiences and try to impose them on the rest of the world in their art.

In our third possible usage of the term there is also a perception that the world has changed since the early years of this century. In the developed ('late capitalist') countries the advances of the communications and electronics industries have (it is argued) revolutionized human society. Instead of reacting to these changes in what is characterized as a Luddite manner, the postmodernist may instead counsel celebration of the present: enjoyment of that loss of artistic aura or 'holiness' which follows what Walter Benjamin (one of the most important theorists of modernism and to a certain extent also a prophet of postmodernism) called 'mechanical reproduction'. In common with some much earlier avant-gardists, many postmodernists are fascinated with rather than repelled by technology, do not reject 'the popular' or the commercial as beneath them, and are very much concerned with the immediate effect of their works: publication is for them (allegedly) more a strategic act than a bid for immortality.

The following writers and their works have been categorized as postmodernist: John Barth, Richard Brautigan, Angela Carter, Thomas Pynchon, Donald Barthelme, William Burroughs, Walter Abish, Alain Robbe-Grillet, Salman Rushdie, Ronald Sukenick, and Jorge Luis Borges.

Finally, just to complicate matters even further, the term *magic realism* should be recalled: see p. 64.

---

[11] See for example David Harvey, *The Condition of Postmodernity* (Oxford, Blackwell, 1989).

6 Analysing Fiction

Narrative technique

1 *Narrators*

In my opening chapter I drew attention to the fact that everything we read in a novel comes to us via some sort of 'telling'. We are told what happens in a novel; no matter how successful the novelist is in making a scene seem dramatic it is never dramatic in the way that a play or a film is. We may feel that we 'see', but we see as a result of what we visualize in response to a narrative not an enactment. Even in those relatively rare cases in which a novelist makes extended use of the present tense, a technique which gives an added sense of immediacy to the narrative, we are still *told* what is happening rather than witnessing it directly as we can with a play or a film. The fact that when reading a novel we know that we can flip forward a page or a chapter, or look at the last page, is thus worth thinking about: it explains that sense we have that in reading a novel we are going through what *has already happened*, that which is being *re*counted to us.

However in one respect the writing of a novel is comparable to the making of a film. When we watch a film we seem to be seeing 'things as they are' – 'reality'. But a director has *chosen how* we see these things, this reality; he or she has decided whether the camera will be placed high or low, whether there will be rapid cuts from one camera angle to another or not, whether a camera will follow one character as he or she walks along a street – and so on. One scene in a film could be shot innumerable ways, and each of these ways would produce a different effect upon the audience. Even with a simple conversation between two characters the audience's attitude towards each character can be affected by different camera angles, cutting, and so on.

The novelist has a far greater range of choices open to him or her than does the film director, and we conventionally refer to that particular selection which he or she makes as the *narrative technique* of a particular work. Narrative technique includes such matters as the choice of narrator and narrative situation, selection and variation of perspective and voice (or 'point of view'), implied narrative *medium*, linguistic register (for example, the choice between colloquial or formal language), and techniques such as Free Indirect Discourse. All of these I will discuss below.

Let us start with narrators, with the individuals, voices, or whatever who (or which!) *tell us* the story. Perhaps most obviously, an author can have the story told through the mediation of a *personified narrator*, a 'teller' recognized by the reader as a distinct person with well-defined individual human characteristics. Alternatively, the narrative source can seem so undefined as to make it doubtful whether or not we are dealing with an individualized human source which comes between the author and the reader. (There is no ideal term to describe such a narrator: *authorial, impersonal* and *third-person* all have their drawbacks. The first suggests identification with the real-life author, the second suggests a lack of intimacy which may be misleading, and the third excludes those first-person narrators who we are unhappy to refer to as 'personified'.)

Consider the narrator of Sterne's *Tristram Shandy*, whose distinct personality is thrust at us in the opening words of the novel:

> I wish either my father or my mother, or indeed both of them, as they were in duty both equally bound to it, had minded what they were about when they begot me . . .

The opening of D.H. Lawrence's *Sons and Lovers* (1913), in contrast, strikes us as far more impersonal; whereas the opening of Sterne's novel focuses our attention on to the narrator, who is talking about himself, the narrative of *Sons and Lovers* focuses our attention on to what is told rather than who tells or how:

> 'The Bottoms' succeeded to 'Hell Row'. Hell Row was a block of thatched, bulging cottages that stood by the brookside on Greenhill Lane. There lived the colliers who worked in the little gin-pits two fields away.

Some narrators may even have names and detailed personal histories, as does Nick Carraway, the narrator of F. Scott Fitzgerald's *The Great Gatsby* (1925). Other narrators merely indicate to us that they are persons – perhaps by the occasional use of 'I' in their narrative – but tell us no more about themselves than this. We thus have a continuum of possibility: (i) personified, named, and with a full human identity; (ii) human but anonymous; (iii) not fully comparable with any human perspective.

Of course one obvious distinction between the narrative of *Tristram Shandy* and that of *Sons and Lovers* is that the first of these novels is narrated in the first person, while the second is what is often referred to as a third-person narrative – meaning a narrative told from a source external to the world of the novel and generally unpersonified. The crucial test here is not just whether or not a narrator refers to him- or herself as 'I' (although this is significant), but whether this 'I' is the (or a) main participant in the created world of the work in question. Third-person narratives are often described as works with omniscient narrators or works characterized by an omniscient point of view. Omniscient means 'all-knowing', and as human beings are not generally all-knowing it is conventional for omniscient narrators to be unpersonified, although this convention is not infrequently broken. The term 'omniscient' is often used in a loose way to indicate any work in which the narrative has access to that which – like a character's secret thoughts – is normally concealed to observers in the real world. Thus a so-called omniscient narrator may not, actually, be completely all-knowing. The opening of Conrad's *Lord Jim*, for example, before the personified narrator Marlow takes over the telling, is told by an unpersonified third-person narrator who is often described as omniscient by critics. But the first sentence of the novel reads, 'He was an inch, perhaps two, under six feet, powerfully built, and he advanced straight at you with a slight stoop of the shoulders, head forward, and a fixed from-under stare which made you think of a charging bull'. The word 'perhaps' reveals that however much this narrator may appear to 'see all', it is not quite all that is always seen.[1] It should be remembered that just as novelists may decide to give a narrator more knowledge than is possessed by an ordinary human being, so too they may decide to restrict this knowledge when it suits them. Complete omniscience is not only unfamiliar to human beings; it may work against the creation of that tension and uncertainty that exercise the reader's mind in a creative fashion.

Thus my earlier comments should not be taken to suggest that if in the course of a long novel the narrative lapses into the first person on one or more occasions then the reader is justified in assuming that the whole novel is told to him or her by the 'person' who lies behind the use of 'I' or 'me'. Of course, were we not dealing with fiction, then this would be

---

[1] One explanation is that Conrad wants the reader to respond to the narrator as to a personified and non-omniscient story-teller at this stage of the novel, while retaining the ability to allow this story-teller omniscience at a later stage. Another explanation is that Conrad is writing against the grain of his inspiration here, and that the narrative really needed a human narrator from its start, and not just from the point at which the personified Marlow takes over.

so: if we were studying a letter from a government department, even a single use of 'I' would justify our assuming that one person had written the letter and might be prepared to take responsibility for it if we were able to trace him or her. But the same is not necessarily true about a novel. A novelist may wish to attribute his or her narrative to a single, personified source at one point in its unfolding, while creating the impression at other such points that it emanates from a more diffuse, less specific or human source. (In his study of Conrad's narrative technique, Jakob Lothe has discussed two such examples of the sudden intrusion of the first person into a previously unpersonified narrative, in *The Secret Agent* and *Nostromo*.)[2] One of the things at which the student of the novel has to become adept is recognizing when such a shift of overt or implied narrative source takes place in a work of fiction. Such shifts can be either marked or unmarked; that is to say, the author may or may not draw attention to them.

In Henry James's *The Turn of the Screw* for example we have a complex narrative structure; an unnamed, 'outer' narrator reports on a scene during which another character named Douglas undertakes to read a story told to him years before by an unnamed governess of his, a written copy of which he has been given by her. The outer narrator then describes Douglas's introductory comments and gives us some comments upon his reading of the manuscript, and then proceeds to reproduce that manuscript for the benefit of readers.

(It is interesting to note that many readers and critics assume that the outer narrator [or 'frame narrator'] is a man – perhaps because 'he' makes many dismissive remarks about the ladies present – but there is no direct evidence that this is the case.[3] Christine Brooke-Rose, in her book *A Rhetoric of the Unreal* [1981] takes many critics of *The Turn of the Screw* to task for reading details in to the work that are not there, but she herself refers to the outer narrator as 'he'. The example shows how *active* the reader inevitably is in creating an image of a narrator on the basis of hints and suggestions in the text. A skilled author will, of course, make use of this fact – sometimes relying upon it, and sometimes challenging and undermining it. We assume, for example, that a personified narrator will

---

[2] Jakob Lothe, *Conrad's Narrative Method* (Oxford, Clarendon Press, 1989).

[3] Since writing this I have discovered that in 1965 A.W. Thomson argued that James's frame narrator was 'fairly obviously a woman'. See his article '*The Turn of the Screw*: Some Points on the Hallucination Theory', *Review of English Studies* VI(4), 1965, pp. 26–36.

belong to one of two genders and will not change sex, but an author can frustrate this assumption if he or she decides that this will be productive.) We have a similar combination of an outer, unnamed narrator and a named 'inner' narrator in Joseph Conrad's *Heart of Darkness*, although here we have no written document that is reproduced, but rather the tale told to a small group by Marlow (the 'inner' narrator) is given to the reader via the anonymous outer narrator.

There is much that can be said about both James's and Conrad's narrative technique, but for the time being I would like to comment upon the fact that both writers feel the need for a combination of named and unnamed narrators. (Whether James's Douglas can, accurately, be described as a narrator is a matter for debate: he certainly reads aloud the governess's story, but his reading does not function as an element in the mediation of the story from assumed source to the implied reader.) The granting of a name conventionally intensifies the degree of personification, the extent to which we think of a narrative source as human and individual, although some unnamed narrators (Jane Austen's, for example) can appear to the reader as highly individual and possessed of something like a personality. (Or is it, rather, that we have a detailed knowledge of the values and attitudes relating to Austen's narratives, without necessarily individualizing or personifying them? Or do we rather attribute them to the person of the author herself?)

If we turn to Charles Dickens's novel *Bleak House* (1852–3), and focus upon those parts of it which are not told by the personified narrator Esther Summerson, we see something very different. Chapter 20 opens thus:

> The long vacation saunters on towards term-time, like an idle river very leisurely strolling down a flat country to the sea. Mr Guppy saunters along with it congenially. He has blunted the blade of his pen-knife, and broken the point off, by sticking that instrument into his desk in every direction. Not that he bears the desk any ill will, but he must do something . . .

This narrative seems human perhaps in its gentle irony, but its viewpoint corresponds to no possible human viewpoint in ordinary life. The narrator is uninvolved dramatically in the scene, but he (is it he?) knows what is going on in Mr Guppy's head perhaps better than does Mr Guppy. Yet at other points in this narrative the narrator betrays an ignorance of things that we have learned from Esther Summerson's narrative. (So the narrator certainly cannot be Charles Dickens himself!)

Dickens uses the present tense in this extract; his anonymous narrator in *Bleak House* makes recurrent use of this tense, and it has a very definite effect on the way the reader responds to the narrative. We often use the present tense to tell stories or jokes, and so one effect of Dickens's use of it here is to make the narrative seem more familiar, intimate, colloquial. At the same time, the use of the present tense gives the scene more *dramatic* force: we feel that we are actually watching Mr Guppy as he is doing something, although without losing a sense of the guiding presence of the narrator. The main point I want to make here, however, is that although there are situations outside of literature in which we deliver present-tense narratives, the extract I have quoted could be nothing other than literary. If we suddenly turned the radio on and heard these words being read, or if we discovered them on a scrap of paper somewhere, we would still know that they had to come from a piece of fiction, for in no other situation would this content and this narrative style be found in such combination.

The narrator, if personified, can have a range of different sorts of relationship with the actions and events described in a novel. He or she may be an 'intra-fictional narrator' – an actor with a full part to play in the story told, or an observer of events in which he or she is personally uninvolved, or an 'extra-fictional narrator' – merely a narrator, telling a story but indicating no personal involvement in or relationship to this story, which may even be presumed to take place on another level of reality from that in which he or she exists.

In Joseph Conrad's *Heart of Darkness*, for instance, his narrator Marlow tells a story in which he personally has been involved. The 'outer', unnamed narrator in this same novella is also in a sense involved in the outer action (the 'frame') of the story, but clearly this narrator's relationship to events in the work is very different from Marlow's. In Charles Dickens's *Great Expectations* (1860–61) the narrator Pip is telling us his life story, but the great difference of age and maturity between the narrating and the narrated Pip means that the narrator can be either very involved or relatively uninvolved in the story at different times. Indeed, when we are talking about fictional narratives such as this it is a good idea to shake off the habit of treating the mature narrating person and the youthful narrated person as the same, and to see them *in narrative terms* as essentially two separate people.

In *Wuthering Heights* Mr Lockwood is in a sense involved in present-time events in the novel (and the novel's time sequence is complicated enough to give us a number of different 'present times'), but he is really an observer of the key events in the work which are narrated by Nelly

Dean to him. And, finally, the narrator of the interchapters in Henry Fielding's *Tom Jones* exists on a different plane from the characters and events in the rest of the story, upon the fictive nature of which he comments directly.

Some theorists use the term 'homodiegetic' to refer to a narrator such as Lockwood who is a narrator of a novel's main story but (and this is the key point) who also participates in the story as a narratee (he listens to Nelly Dean's narrative, which is directed at him). A further term you may encounter is 'autodiegetic': an autodiegetic narrator is him- or herself the main character in the story he or she tells (thus in her own narrative, the governess in *The Turn of the Screw* plays an autodiegetic rôle).

2 *Narrative medium and language*

In addition to choosing a narrator or narrative source, the novelist can (but need not) select a stated or implied *medium* for his or her narrative. Now of course all novels consist of written, normally printed, words in a literal sense, but a novel may well be presented to the reader as if it were spoken rather than written or thought, or it may even be presented in such a way as to suggest that it is a sort of medium-less narrative – something impossible outside the realms of literature.

Thus, to take two very different examples, Agatha Christie's *The Murder of Roger Ackroyd* (1926) and Feodor Dostoyevsky's *Notes from Underground* (1864) are both presented to the reader as *written* documents. We learn this only towards the end of Christie's story, when the narrator explains how he came to write his account (and that should make us pause for thought: think of the oddness of reading a story and not knowing until the end of the story whether it is a written document we are reading: that is because the written text *we* see stands for another text, about the precise nature of which we can be left in the dark by the novelist). In Dostoyevsky's story we find the following interpolated comment on the first page:

> (A poor witticism; but I won't cross it out. When I wrote that down, I thought it would seem very pointed: now, when I see that I was simply trying to be clever and cynical, I shall leave it in on purpose.)

We may well ask: why should one author want to reveal the implied narrative medium of the story on the first page, and another only towards the end? What difference does it make to the reader's response to the story? (Compare Philip Roth's *Portnoy's Complaint* [1969], in which the fact that the narrative is a spoken one is revealed on the last page along with the precise context of its delivery.)

Where a novelist does not state his or her implied narrative medium, all sorts of variations are possible. Emily Brontë's *Wuthering Heights* opens as if it might be some sort of diary entry:

> 1801. – I have just returned from a visit to my landlord – the solitary neighbour that I shall be troubled with. This is certainly a beautiful country!

But other parts of the novel, also told by Mr Lockwood as is this opening passage, read more like thought-processes than writing. This is the third paragraph of Chapter 10:

> This is quite an easy interval. I am too weak to read, yet I feel as if I could enjoy something interesting. Why not have up Mrs Dean to finish her tale? I can recollect its chief incidents, as far as she had gone. Yes, I remember her hero had run off, and never been heard of for three years: and the heroine was married: I'll ring; she'll be delighted to find me capable of talking cheerfully.
>
> Mrs Dean came.

There is, I think, a slight clumsiness here, which may be attributed to the fact that novelists following Brontë perfected techniques for making such transitions between narrative viewpoints and media less obtrusive.

Closely related to the foregoing issues is the matter of *linguistic register*. Whether a narrative is formal or colloquial, for example, depends a lot upon whether it is told by a personified or unpersonified narrator, and upon whether we are to assume that it is spoken, written, or thought. (Conversely, of course, the degree of informality of a novel's language may lead us to infer that a narrative is, for example, a spoken tale by a personified narrator.) Language in one sense is the medium of a novel as paint is the medium of a painting, but this is a very bad comparison in general as painters do not generally represent paint whereas novelists do often represent language: *Heart of Darkness,* for example, is not just *in* language, it *represents language* – the language uttered by Marlow and used by the novella's frame narrator. Thus in one sense the *actual* medium of this novel – language – is also its *implied* medium. (Although of course the implied medium of *Heart of Darkness* is spoken, not written, language.)

Thus while a novel's actual medium is written language its implied medium need not be, and indeed as I have already pointed out a novel need have no implied medium at all, and the text of a novel need not even represent language, it could represent a state of mind, a sequence of events and experiences that have not necessarily been verbalized by anyone but which are translated into words for the reader by the novelist, and so on.

Some narrative media can be entirely unspecific. Joseph Conrad's *The Shadow-Line* (1916–17), for example, can be read as if written, spoken or thought in origin; it is indeed probably wrong to try to fix a medium for it as it represents a sort of medium-less, pure narrative which we take in as readers without worrying about its implied physical origin. A number of recent theorists have insisted that much narrative in the novel is of this sort; unlike non-fictional narratives, the telling of a novel need not involve a physical medium.

3 Narrative and representation

This may seem to contradict what I said earlier about all narrative being a telling: does not a telling necessarily involve a medium by means of which the telling takes place? It should be said that this touches upon a number of very complex theoretical disputes in the theory of narrative. Some theorists prefer to distinguish between representation and telling, and claim that the novelist can represent – say – a character's consciousness in such a way as gives the reader knowledge of this consciousness without any sense of having received this through a telling – something that is, of course, impossible outside the realms of written narrative. Thus the narrative theorist Ann Banfield has argued that representation is, as the name suggests, a 're-presentation' which makes 'present' what is either absent or past, and which is thus unlike narration, having only a single point of reference with respect to time and place.[4] (A narration has two: the time and place of the narrated events, and the time and place of their narrating.) Banfield thus argues that so far as certain examples of representation are concerned, to inquire into the process of 'telling' bespeaks only the critic's lack of comprehension of what representation involves, hence the title of the book in which she argues this case – *Unspeakable Sentences*. In Banfield's view the novelist may give us

[4] Ann Banfield, *Unspeakable Sentences: Narration and Representation in the Language of Fiction* (London, Routledge, 1982), p. 268.

sentences which are not and cannot be spoken by anyone; their function is that of representation not telling or narrating.

Banfield's views are controversial, but at the very least they should remind us that whereas in the extra-fictional world any statement must come from some source, in fiction certain techniques may be designed to allow the reader to experience events, states of mind, dispositions, which are not expressed or communicated but which just *are*. The fact that the novelist succeeds in communicating these to us should not obscure the point being made: when we read a novel we may be given the wherewithal to undergo states which (as we know from our own subjective lives) are not expressed or told, but which are nonetheless lived through.

4 *Ways of telling*

Let us pause for a moment and ask why all this is important. What difference does it make? Well, in everyday life *who* tells us a story, and *how,* make a very big difference. The statement that the economy has never been stronger has a different effect on us if told by the Prime Minister from the effect it has on us if we read it in an opposition newspaper. A proposal of marriage uttered directly, in emotional speech, strikes the recipient rather differently from one formally written and received by post. Source and medium affect the *selection*, the *authority*, and the *attitude towards what is recounted* of the narrative – and thus, of course, the effect on the reader or listener. (And one of the problems involved in Banfield's view of representation is, it seems to me, that even such representation involves selection, and the act of selection implies some form of mediation.)

The same is true with the novel: different narrators, different narrative media *change* a story; they affect not just how we are told something but what we are told, and what attitude we take towards what we are told. Nobody knows this better than novelists themselves, and different novelists have given us some revealing accounts of the problematic processes that deciding upon a narrative technique involves. There is, for example, a fascinating entry in Virginia Woolf's diary dated 25 September 1929, outlining her plans for her novel *The Waves*, which at this stage of her work on it she had provisionally entitled *The Moths*. She writes:

> Yesterday morning I made another start on *The Moths*, but that won't be its title; and several problems cry out at once to be solved. Who thinks

it? And am I outside the thinker? One wants some device which is not a trick.[5]

Note how the choosing of a particular perspective, a particular centre of perception for the telling, presents itself as a *problem* right at the start of Woolf's work on this novel. Note too her distinction between 'I' and 'the thinker' – in other words, between (I think) the creative-moral centre of authority, and the perceiving-observing-telling situation, in the novel. It is very important to recognize the fact these two are not necessarily identical – either in *The Waves* or in any other novel or short story. And finally, that comment that she 'wants some device which is not a trick' also underlines the fact that this sort of choice is not a purely technical one, it has aesthetic and moral implications which demand careful consideration. (Woolf was capable of lighter and more humorous writing in which narrative 'tricks' played their part: thus in *Orlando* her heroine changes sex in the course of the work, while her biography *Flush* [1933] is written from the perspective of Elizabeth Barrett Browning's pet spaniel. In both cases, of course, the 'tricks' had their serious sides.)

A succession of fascinating letters written by Jean Rhys provides relevant evidence concerning her choice of a way of telling[6] her novel *Wide Sargasso Sea*. (In Chapter 2 I discuss the complex intertextual relationship between this novel and Charlotte Brontë's *Jane Eyre*.) To her daughter, in 1959, Jean Rhys writes about the forthcoming novel:

> It can be done 3 ways. (1) Straight. Childhood. Marriage, Finale told in 1st person. Or it can be done (2) Man's point of view (3) Woman's ditto both 1st person. Or it can be told in the third person with the writer as the Almighty. . . .
> I am doing (2).

Later on in the same year she writes to Francis Wyndham admitting that the novel had already been written three times, first told by the heroine in the first person, second told by the housekeeper Mrs Poole as the 'I', and now using two 'I's: 'Mr Rochester and his first wife'. According to Rhys

---

[5] Leonard Woolf (ed.), *A Writer's Diary: Being Extracts From the Diary of Virginia Woolf* (London, Hogarth Press, 1972), p. 146.

[6] 'Way of telling' is a formulation which includes both 'perspective' and 'voice'. For this distinction, see p. 103. What Rhys is concerned about here is, strictly speaking, the choice of a particular voice, although this has implications for narrative perspective.

it was unsatisfactory to have the heroine tell all the story herself because the result was obscure and 'all on one note', and it was also unsatisfactory to have the housekeeper Grace Poole tell the whole story because although technically this was the best solution the character 'wouldn't come to life'.[7] Before we leave this example, you should think a bit about the phrase 'Man's point of view'. Note how this has both a technical and an ideological or 'sexual political' meaning: a writer's choice of 'point of view' is rarely just a technical matter. Note too that it can be a matter of trial and error for a novelist to hit on the right way of telling for a particular story.

Other considerations can also be important. Christopher Isherwood's autobiography *Christopher and his Kind* contains an enthralling account of his problems in fixing the narrative voice for his novel *Mr Norris Changes Trains* (1935). He felt that he wanted the reader to experience Arthur Norris personally, and believed that this could only be done by writing in the first person, subjectively, and having the narrator meet Norris. But the narrator could not be Christopher Isherwood because Isherwood was not prepared to reveal his own homosexuality publicly by having the narrator an avowed homosexual, although he scorned to have the narrator heterosexual. Thus the narrator of the work ends up with no explicit sex-experiences in the story, and was dubbed a 'sexless nitwit' by one reviewer. But an additional interesting aspect of the problem is mentioned by Isherwood. He wanted the reader's attention to be concentrated upon Norris and not on the narrator, and if the narrator had been made a homosexual he would have become so 'odd' and interesting that he would have thrown the novel out of balance by attracting too much of the reader's attention.

Isherwood makes a further interesting point. He notes that if a narrator is given no personal qualities then when something happens to him the reader will assume that his responses are not likely to be unusual; the reader can thus identify with the narrator and experience with and through him. But a narrator who is somehow 'odd' will prevent the reader from identifying with him even if he or she is sympathetic to him.[8]

Let us return to the issue of choosing a way of telling in the light of our understanding of the crucial importance of such decisions by the

---

[7] Francis Wyndham and Diana Melly (eds), *Jean Rhys: Letters 1931–1966* (London, André Deutsch, 1984), p. 162.

[8] Christopher Isherwood, *Christopher and his Kind: 1929–1939* (London, Methuen, 1977), pp. 141–2.

novelist. Many critics have found it useful to distinguish between *reliable* and *unreliable* narrators, a distinction that touches upon some of the points raised by Isherwood. We can also note that, reliability apart, we associate some narrative choices more with the views and position of actual authors, and some not at all (or far less) with their creators. In general we can say that a single, consistent, unpersonified voice is more likely to be associated with authorial beliefs than is a personified narrator in a novel with many narrators, although of course in both cases this depends upon the attitudes expressed in and revealed by the narrative.

Thus it is easier to assume that the opinions expressed by the anonymous narrator of Conrad's *The Secret Agent* (1907) are close to those of Conrad himself than it is in the case of his personified narrator Marlow, who appears in a number of his works. Moreover, even if Marlow is a generally reliable narrator, we do not take everything he says on quite the same trust as we do what the narrator of *The Secret Agent* tells us, simply because we see Marlow from the outside and treat him as another person, whereas we adopt the perspective of the narrator of *The Secret Agent* and, as we read, think ourselves fully into his position from the inside.

Consistency is a crucial issue here. An inconsistent narrator cannot, logically, be wholly reliable, although we may recognize in fiction as in life that inconsistency may be the result of a continued and painful attempt to be truthful and accurate. The fact that Gulliver in Swift's *Gulliver's Travels* seems to vary from book to book, being alternatively percipient and obtuse, blindly patriotic and unchauvinistically humanistic, warns us that we can relax into no unguarded acceptance of his statements or opinions. On the other hand, although few if any readers of *Wuthering Heights* can identify totally with Mr Lockwood, he is consistently portrayed, and so we feel more and more confident at assessing his opinions in the light of our view of his personality and character.

In addition to choosing a narrator and (perhaps) an implied narrative medium, the novelist has to select a *form of address* for his narrator. In so doing, he or she normally helps to define what has been termed a *narratee*: that is, the person to whom the narrative is addressed. (Note that the narratee may be the actual reader, but is not necessarily so. The narrative can, for instance, be directly addressed to the 'Dear Reader', or it can be spoken or written to another intra-fictional[9] target or destina-

---

[9] I now prefer to indicate whether or not something or someone belongs to the 'fictional world' of a novel by means of the terms 'intra-fictional' and 'extra-fictional', as I find these less ambiguous and more immediately understandable

tion, or it can be projected into a void. In his novel *Tom Jones* Henry Fielding uses 'interchapters' in which the reader is addressed directly, while the remainder of the novel is less overtly pointed at any reader or listener.)

What, for example, do we make of the opening lines of Albert Camus's *The Outsider* (1942)?

> Mother died today. Or, maybe, yesterday; I can't be sure. The telegram from the Home says: *Your mother passed away. Funeral tomorrow. Deep sympathy.* Which leaves the matter doubtful: it could have been yesterday.

Clearly this could be addressed to someone (would there be any sense in saying it to oneself?), but the narrative might represent a state of mind and a sequence of events which the reader is not meant to think are actually aimed at any recipient in particular.

The same fictional character can be both narrator and narratee: in the final section of James Joyce's *Ulysses*, for example, Molly Bloom is both the narrator and the narratee. Part of the narrative complexity of *The Turn of the Screw* comes from the fact that there are different narratees for different levels of narrative. The frame narrative is aimed at no specified narratee, but nonetheless gives the sense at being aimed at a human destination within the world of the telling (i.e. not the actual reader of the tale); the governess's narrative has Douglas as its narratee when it is told to him, and perhaps also in its written version (but perhaps not: we do not know when she writes her story, and it may even have been written before she meets Douglas). And Douglas's comments about the story in the opening pages of *A Turn of the Screw* have those listening to him as the intended target, so that they are also in a certain sense narratees. One of the revealing things about this work is that a piece of prose fiction which appears relatively simple, whose narrative technique involves the common reader in no problems of understanding or interpretation, can nevertheless be seen to involve enormously complex narrative mediations when analysed.

Narratives can also involve such elements as *complicity*, *intrusion*, and *intimacy* – all of which are normally instantly recognized by readers while remaining tricky to analyse.

than the terms 'intradiegetic' and 'extradiegetic', which are often used by writers about fiction.

Take for example the following brief extract from Chapter 3 of E.M. Forster's *Howards End*, in which Mrs Munt is talking at cross-purposes to Charles Wilcox, believing him to be engaged to her niece Helen:

'This is very good of you,' said Mrs Munt, as she settled into a luxurious cavern of red leather, and suffered her person to be padded with rugs and shawls. She was more civil than she had intended, but really this young man was very kind. Moreover, she was a little afraid of him: his self-possession was extraordinary. 'Very good indeed,' she repeated, adding: 'It is just what I should have wished.'

'Very good of you to say so,' he replied, with a slight look of surprise, which, like most slight looks, escaped Mrs Munt's attention.

That final narrative comment is the culminating stroke in a process whereby the reader is sucked into complicity with the narrator. We are amused with the narrator at Mrs Munt's obtuseness and self-importance, and as a result of such passages we are likely to be far more malleable in the hands of the narrator, far more willing to accept his value-judgements and assessments of characters. It should perhaps be added that at this and other points many readers feel that it is the author himself with whom we become complicit, and although this view has become unfashionable in recent decades it should not be dismissed. (It is perhaps for this reason that it seems natural to refer to this narrator as 'he'.)

If we think of Jane Austen's intimate address to the reader in her novels, and Henry Fielding's intrusive interpolation of interchapters in *Tom Jones*, we can see that the narrator has a wide variety of relationships with the narrated events and characters, and with the reader, at his or her disposal. Choosing the right relationships is essential to success for the writer of fiction. For many readers, the last paragraph of Thomas Hardy's *Tess of the d'Urbervilles* (1891) starts with an annoyingly intrusive and heavy-handed comment from the narrator, a comment which such readers feel detracts from the power of the scene depicted at this point of the novel: 'Justice was done, and the President of the Immortals, in Aeschylean phrase, had ended his sport with Tess.' In contrast, at the end of the first chapter of Jane Austen's *Mansfield Park* we find this concluding sentence about Mrs Price (Fanny's mother): 'Poor woman! she probably thought change of air might agree with many of her children.' Although technically intrusive, this seems less to break in upon an established tone or perspective than does Hardy's comment, and so seems to have offended fewer readers than has that final paragraph of *Tess of the d'Urbervilles*.

It will be understood that the word 'intrusion' suggests that the fictional world is to a certain extent closed off from the real world of the author or even of his or her narrator. But, as my earlier comments on modernism and postmodernism have I hope suggested, much fiction that can be classified as modernist or postmodernist does not present the reader with a separate, enclosed, fictional world. Instead, it is characterized by a breaking-down of that firm distinction between the fictional and the real that we associate with classical realism. One of the paradoxes about classical realism is that it rarely admits to its own fictionality – something to which modernist and postmodernist fiction confesses most frequently. This is paradoxical, because it means that the classical realist novel wants to portray everything in the world as it is with the exception of itself! Thus it is no accident that I have chosen to exemplify narrative intrusion by reference to novels by Jane Austen and E.M. Forster – both writers of works which, albeit in very different ways, are more satisfactorily classified as realist than as modernist or postmodernist. Interjections from the real or implied author in much modernist and postmodernist fiction are so frequent as hardly to represent anything that the reader experiences as intrusion.

A narrative can be either *recollective* or *dramatic*. Of course, *any* narrative involves recollection; to be told a story is to be informed of something which has already happened, something which is being remembered, recounted. We should probably except those rare examples of future narrative, although even here, because we are given the impression that the narrative *knows* what events are to happen, in a sense they exist already in completed form – have already taken place and are being recollected. In Margaret Atwood's short story 'Weight' (1991), for example, the present-tense, first-person narrative gives way in the four final paragraphs to the future tense. But in these paragraphs the speaker is thinking about what she will do the next day and what will happen then, and so the emphasis is still on the present thinking rather than the future happening.

Present-tense narrative is more problematic, but it can be argued that this technique at least implies a gap between the happening and the recounting, although in English the use of the present continuous can give a sense of immediacy. Thus on the penultimate page of Atwood's 'Weight' we can read the following paragraph:

> Charles is walking me to the door, past white tablecloth after white tablecloth, each held in place by at least four pin-striped elbows. It's like the *Titanic* just before the iceberg: power and influence disporting

themselves, not a care in the world. What do they know about the serfs in steerage? Piss all, and pass the port.[10]

There is no doubt that the reader is given a sense of being present as things happen here, although I think that probably this is a sense of the female narrator's retrospective imagining-what-it-was-like-when-it-took -place: the things that are happening are unfolding in her head as she remembers the events in the restaurant, they are not the actual events.

Such subtleties and marginal cases should not obscure what every sensitive reader of fiction knows: some ways of telling a story can have far more dramatic effect than others (as is clear from Samuel Richardson's earlier-quoted comment on his epistolary technique). Note how Jean Rhys changes the tense used in the following passage in order to create dramatic immediacy in her novel *Good Morning, Midnight* (1939):

> The lavatory at the station – that was the next time I cried. I had just been sick. I was so afraid I might be going to have a baby. . . .
> Although I have been so sick, I don't feel any better. I lean up against the wall, icy cold and sweating. Someone tries the door, and I pull myself together, stop crying and powder my face. (Ellipsis in original)

We can contrast a passage from a much-quoted scene in Chapter 4 of Charles Dickens's *Great Expectations*. The young Pip has just robbed the pantry of food which he has given to the escaped convict who has terrified him, and is eating Christmas dinner with his family in fear and trembling:

> Among this good company I should have felt myself, even if I hadn't robbed the pantry, in a false position. Not because I was squeezed in at an acute angle of the table-cloth, with the table in my chest, and the Pumblechookian elbow in my eye, nor because I was not allowed to speak (I didn't want to speak), nor because I was regaled with the scaly tips of the drumsticks of the fowls, and with those obscure corners of pork of which the pig, when living, had had the least reason to be vain. No; I should not have minded that, if they would only have left me alone. But they wouldn't leave me alone.

---

[10] Margaret Atwood, 'Weight'. In *Wilderness Tips* (London, Bloomsbury, 1991), p. 192.

Here the author does all in his power to reduce immediacy and to *distance* the reader from the child's experiences. In reality a young child in Pip's situation would have suffered torments, and would certainly not have found anything amusing in the experiences he was undergoing. But the reader is not encouraged to experience with the young Pip: rather we look at him from the outside, experiencing rather with the older, narrating Pip and sharing his humorous view of the events from his childhood. Such narrative distancing does not necessarily prevent the reader from experiencing with a character – it can sometimes involve deliberate understatement which allows the reader to imagine for him- or herself what the experience involved must have been like. But generally speaking a novelist can control the extent to which a reader empathizes with a character by means of the manipulation of narrative distance. The more 'here and now' the narrative is – both spatially and temporally – the more likely the reader is to enter into the character's experiences.

We can sum up much of the force of the discussion so far in this chapter with the question: 'What does the narrative know?' Literary critics have traditionally used the technical term *point of view* to pinpoint this particular issue. The term has the disadvantage that different critics have used it in slightly different ways, to mean either the *narrator*'s relation to the story told or the *writer*'s attitude to his or her work. This distinction should be borne in mind, but one should also remember of course that an author conveys something about his or her attitude towards the work by choosing a particular relationship to the story told for his or her chosen narrator. Thus in the passage from *Great Expectations* discussed above, the narrating Pip's distance from his youthful self's Christmas dinner agonies suggests that Dickens was not at this point too concerned to explore these agonies in detail.

Another disadvantage with the term 'point of view' is that it obscures what recent theorists have identified as an extremely important distinction, that between *voice* and *perspective*. We can sum up what these terms cover in two questions: 'Who speaks?' and 'Who sees?'. In Katherine Mansfield's short story 'The Voyage' (1922), for instance, the voice is essentially that of a third-person narrator who is semi-omniscient and 'out of the story'. But the reader is, nevertheless, encouraged to see everything through the eyes of the main character of the story, the little girl Fenella. We experience through her senses, even though what we know of these experiences comes to us via a third-person narrative. Hers is the only consciousness we enter (it is always worth asking to whose thoughts we are made privy in a novel or a short story, as the answer gives us significant information about the author's focus of concern). In 'The

Voyage', then, the *voice* is that of a semi-omniscient third-person narrator, but the *perspective* is Fenella's. We can compare a comparable technique much beloved of the popular press. When we read in a newspaper 'We in Britain (or America, or wherever) will not put up with . . .', we are encouraged to read as if we ourselves were speaking through the newspaper. But in effect the writer of such comments really means 'I hope to persuade the British people that . . .'. A knowledge of narrative technique can be useful quite apart from the study of literature.

We should also get used to asking whether the view of characters and events which are given by the narrative is a recognizably human one: in other words, does it resemble that which a real human being might actually have experienced, were the characters and events real? In the case of Conrad's Marlow, Defoe's Moll Flanders, Dickens's Pip or Jean Rhys's Sasha Jensen we can say that a real human being might share such a relationship to the events depicted as they have (which is not to say that such a real human being would have *told* his or her story as they do – or even at all). But no human being in real life would know what the narrator of George Eliot's *Middlemarch* knows about the characters and events in that work.

What difference does this make? Well, a good way to demonstrate this is by means of what we can call the *transposition test*: transposing a novel (in one's imagination) from one 'camera angle' to another. Imagine *Middlemarch* told in the first person by Dorothea, or by Dorothea and Lydgate in successive sections. What would *Wuthering Heights* be like as an epistolary novel, with Isabella's letter about her marriage to Heathcliff as only one of a whole sequence of letters describing the events and developments of the work? Or told exclusively through Heathcliff's eyes? It can be an extremely revealing exercise to take a brief passage from a novel and to rewrite it in this manner.

Thus it is interesting to know that Franz Kafka started his novel *The Castle* (1926) in the first person, and that the manuscript of the novel has 'I', crossed out and replaced by 'K', throughout the early part of the work. If you have read this work you will appreciate how very differently it reads with 'I' rather than 'K'; the reader's whole relationship with the main character is quite different. Kafka clearly found the right voice and perspective for his novel only after having started writing the work – when he could perhaps more easily put himself into the position of a potential reader.

One term which has, I think, a useful summarizing scope here is *narrative situation*. We can indicate what this term covers by asking about the relationship *between the telling and what is told*. Is the narrator personified, if so is he/she a character involved in the action of the work, is the narrative dramatic and immediate or distanced? – and so on. Narrative situation thus includes both perspective and voice, and also a range of other matters such as tone, mood, and frequency (see below for discussion of these terms).

Sometimes attempts to turn novels into plays or films have the effect of demonstrating how crucial their particular narrative situation is. Joseph Conrad attempted dramatizations of a number of his works, all of which were more or less disastrous. I think that we can easily understand that a dramatization of *The Secret Agent* (which Conrad actually completed) *would* be disastrous because it necessarily loses what is at the heart of the work's power: the bitter but pityingly ironic attitude of the narrator towards the characters and events of the novel. (Because plays are not narrative but dramatic in nature they have no narrators – unless the writer is using modernist techniques such as those employed by the German playwright Bertolt Brecht.)

Think of the difference between watching a scene from a film, and being present on the set as it is being made. In a sense we see the same things in both cases, but only in a sense. When we see the scene in the finished film we see the events as the director wants us to see them: we are *situated* in a particular way towards them. What we see on the set is more the raw material, the ingredients from which a skilled director can, by selection and combination, create something meaningful artistically. This is one of the reasons why those involved in real historical events very often tend to be dissatisfied with filmed versions of the same events: to become art the actuality has to be shaped and formed by a director, and this involves selecting only a small part of 'what actually happened' so as to stress certain patterns, certain processes, certain experiences. When a novel is dramatized we still of course have the shaping hand of a director, and the interpretations of the individual actors. But the novel itself gives both far less and far more: the novelist guides our attention to some things and prevents us from becoming concerned with others (the young Pip's real suffering at the Christmas dinner, for example). Much is thus 'cut out'. But what is left is suggestive and artistically rich, such that although a film of a scene from a novel may seem to include much more concrete detail than the novel, it is often more difficult artistically to organize this detail compared to the original arrangement on the printed page. Thus our experience of the film of a novel is often that it seems flatter, less complex, than a reading of the novel itself.

Let me give an example. Charles Dickens's novel *Hard Times* (1854) has been dramatized for television – not unsuccessfully. One quite striking scene in the dramatization I saw many years ago involved Mr Gradgrind's informing his daughter Louisa that Mr Bounderby had proposed marriage to her. The chapter in which this takes place, number 15 in 'Book the First' of the novel, is characterized by the consistently ironic tone that suffuses the whole narrative. The scene is set in Mr Gradgrind's Observatory:

> To this Observatory, then: a stern room, with a deadly statistical clock in it, which measured every second with a beat like a rap on a coffin-lid; Louisa repaired on the appointed morning.

At key moments in the chapter Dickens reminds us of this 'deadly statistical clock' – a perfect symbol of that mechanical measurement, accurate but inhuman, which the whole of *Hard Times* is designed to condemn. For the television director this presented a problem. First, clocks in real life do not start ticking only when you think of them: they tick all the time. In the novel, of course, when no mention is made of the clock it does not occupy our attention. Second, how can a dramatization show us a clock and hope that we will associate it with deadly statistics? The director did his best as I recall, but the effect was significantly less incisive and fine-tuned than what we get on the printed page. As we watched the scene we were made more or less aware of the clock, and some of the associations of inhuman measurement were probably aroused even for the viewer who had not read the novel. But in reading the novel we see the clock through the eyes and with the thoughts of Dickens's narrator. It means what he wants it to mean. And it is the narrative situation that Dickens has created for this scene and for the novel as a whole that makes this possible.

For the purpose of clear exposition I have written so far much as if all novels have one, consistent and unvarying narrative situation. But this is clearly untrue. Again we can make a useful comparison with the cinema. Early films involved one camera mounted rigidly in front of the actors like a theatre audience, before which acting took place. But before long film-makers realized that different cameras could be used and shots from each spliced together to make the film more alive – and that cameras could move with the actors and could utilize long-range and close-up lenses. In this respect a sophisticated novel like *Wuthering Heights* is like a modern film: no longer do we have the single camera of novels such as *Clarissa* or *Robinson Crusoe*, but a range of different, moving recorders. Sometimes such shifts are almost imperceptible. Most readers need to have it pointed out to them after an initial reading of Joseph Conrad's *The*

Nigger of the 'Narcissus' that although the narrative in the work unfolds as if it were consistent, at times the narrator seems omniscient, at times he refers to the crew as 'them', and at yet other times he refers to the crew as 'us'. We can find comparable variations in the narrative of Dostoyevsky's *The Devils* (1870).

It is, in conclusion, important to be able to see narrative techniques in their historical context and development, as well as appreciating the 'internal', technical reasons for developments in narrative technique. The rise of the epistolary novel in the eighteenth century cannot be understood apart from the much greater importance of letter-writing at that time, and the emergence of the stream of consciousness novel in the twentieth century has to be related to the development of modern psychology and the increasing interest in mental operations that accompanies it. The following factors are all important in assessing the significance of a particular narrative technique.

1 Changes in the dominant modes of human communication (think of the enormous effect that the telephone has had on us and that the computer is having).

2 The effect of different world-views, philosophies, and ideologies (there is clearly a parallel between a belief in a God who sees everything, and novelists' use of omniscient narrators – witness Jean Rhys's phrase 'told in the third person with the writer as the Almighty' (p. 96). The widespread loss of belief in such a God seems to have been paralleled by a disenchantment with the possibilities of narrative omniscience).

3 Changes in readership patterns and habits (it is perhaps harder to feel intimate with a larger, more amorphous and anonymous set of readers – or to feel at ease with readers mainly of the opposite sex from oneself).

4 Larger changes in human life and modes of consciousness (think of the growth of urban living, of mass communication, of modern science and politics).

Let me at this point say a few words about *stream of consciousness* technique and the *internal (or interior) monologue*. The two need carefully to be distinguished. An internal monologue necessarily implies the use of *language*, and if an individual is 'talking to himself or herself' then that in turn presupposes a certain amount of consciousness of what is going on in that person's mind. The famous closing section of James Joyce's *Ulysses* gives us, essentially, the stream of Molly Bloom's consciousness, but in internal monologue; she is (I think) thinking to herself in words and is conscious of what she is thinking. I say 'I think' because novelists can transpose unverbalized thoughts and sensations into verbal statements which are attributed to the character in question, and it is possible that at least part of this chapter should be read in this way. Thus when the chapter opens:

> Yes because he never did a thing like that before as ask to get his breakfast in bed with a couple of eggs since the *City Arms* hotel when he used to be pretending to be laid up with a sick voice . . .

it is possible that the reader is meant to imagine Molly's having a primarily sensuous (visual/tactile) memory of the previous recumbent breakfast, and that the words we read are to be taken as a token of this set of sensuous memories. But it seems unlikely – first because the chapter in spite of its lack of punctuation has a logical structure and presents us with Molly's *arguments*, and it is difficult if not impossible to argue non-verbally, and second because Joyce does typically present thought-processes as verbalized.

But consider the following passage – the first two paragraphs of D.H. Lawrence's short story 'Fanny and Annie' (1922):

> Flame-lurid his face as he turned among the throng of flame-lit and dark faces upon the platform. In the light of the furnace she caught sight of his drifting countenance, like a piece of floating fire. And the nostalgia, the doom of homecoming went through her veins like a drug. His eternal face, flame-lit now! The pulse and darkness of red fire from the furnace towers in the sky, lighting the desultory, industrial crowd on the wayside station, lit him and went out.
>
> Of course he did not see her. Flame-lit and unseeing! Always the same, with his meeting eyebrows, his common cap, and his red-and-black scarf knotted round his throat. Not even a collar to meet her! The flames had sunk, there was shadow.

The passage is extremely complex, partly because it mixes what the narrator tells us and what Fanny is thinking, but also because at several points it is hard if not impossible to tell whether what we are reading represents a narrative comment or Fanny's thoughts. I will come back to this point, but for the time being would like to draw attention to the fourth sentence ('His eternal face, flame-lit now!') and the whole of the second paragraph down to the end of the penultimate sentence quoted ('. . . to meet her!'). It seems to me highly likely that these sentences represent Fanny's stream of consciousness rather than the narrator's commentary, although some parts could be both or either. But what seems almost certain is that Fanny is not thinking by means of the actual words we are given. She is not thinking 'His eternal face, flame-lit now! Of course he does not see me, flame-lit and unseeing! Always the same, with his meeting eyebrows, his common cap, and his red-and-black scarf knotted round his throat. Not even a collar to meet me!', although she may verbalize *some* of these thoughts by means of the words we read. What Lawrence's words do is to give us Fanny's stream of consciousness rather than her internal monologue. We can summarize and say that an internal monologue gives us a character's stream of consciousness (or at least part of it, if we imagine images accompanying words), but not every stream of consciousness is an internal monologue.

5 *Free Indirect Discourse*

The difficulty I mentioned above of distinguishing between what a narrator says and what a character thinks or verbalizes may become particularly acute with the use of a technique which is known by a number of different terms: 'Free Indirect Discourse (or Speech)', 'Represented Speech and Thought', 'Narrated Monologue', or the German and French terms, 'Erlebte Rede' and 'Style Indirect Libre'. The complications of terminology are worth wading through, for the technique itself is arguably one of the most important in the novel since Jane Austen. Look again at the opening of 'Fanny and Annie'. Who is it who says 'Of course he did not see her'? The grammatical structure of the sentence, with Fanny referred to as 'her', might suggest that it is the narrator (it cannot be Harry, the man, precisely because he has not seen her). But isn't it rather odd for the narrator to utter this comment, as if Harry's behaviour were expected but nonetheless irritating, and as if the reader knew enough about Harry to legitimize that 'of course'? The same can be said of 'Always the same, with his meeting eyebrows, his common cap, and his red-and-black scarf knotted round his throat. Not even a collar to meet her!' Isn't it odd for the narrator to say this, first because we do not expect an omniscient narrator to get so quickly irritated by a character, and second because the statement assumes that whoever hears this comment will be familiar with poor Harry's shortcomings. How do we know that Harry is always the same? This is the first time that we have encountered him!

In fact, as most readers have no difficulty in working out, it is not Lawrence's narrator but Fanny whose irritation at Harry's familiar inadequacies is thereby expressed. Typical of Free Indirect Discourse – to use what I now favour as the best term to describe this technique – is the use of the grammar of third-person utterance (with certain modifications) to present us with a character's speech or (verbal or non-verbal) thoughts. (As will be understood, terms such as Free Indirect Speech, represented speech, and represented thought have a more limited application than does the blanket-term Free Indirect Discourse.)

The traditional way of defining what we can refer to as FID makes use of grammatical or linguistic evidence. This involves seeing FID as a midway point between Direct, and Indirect (or Reported) Discourse (DD and ID), or as a combination of the two which combines their grammatical characteristics in a distinctive mix. Thus we can use the following simple illustration, in which one can note that the third example retains the third person 'he' and past tense from ID, but in its truncation resembles the words in inverted commas in the DD example.

DD: He said, 'I love her'
ID: He said that he loved her
FID: He loved her[11]

What should strike us immediately about the third example ('He loved her') is that, like sentences in the passage from 'Fanny and Annie' at which we have already looked, in the context of a passage from a novel it could be ambiguous. It could either give us the consciousness of the 'he' referred to, his realization that he loves a woman, or it could serve as the statement of a narrator about a character (who might not even realize, yet, that he loves the person concerned). This ambiguity is exploited by the most skilled writers of prose fiction, and very often allows them to move backwards and forwards between narrative comment and character consciousness, often with no apparent seams. Look again at the opening of Lawrence's story. It clearly gives us both comment that comes from a narrator, or source outside the depicted characters and events, and also what Fanny is thinking. But it is impossible to make a sharp distinction between the two such that one can end up with two neat piles, one labelled 'narrator's comment' and the other 'Fanny's thoughts'. This, I would argue, is a strength rather than a weakness of the passage; if you disagree, try rewriting it so as to exclude any ambiguity as to whom statements should be attributed.

Another example, this time a passage from Katherine Mansfield's 'The Voyage':

[11] Example taken from Shlomith Rimmon-Kenan, *Narrative Fiction: Contemporary Poetics* (London, Methuen, 1983), p. 111.

'How long am I going to stay?' she whispered anxiously. He wouldn't look at her. He shook her off gently, and gently said, 'We'll see about that. Here! Where's your hand?' He pressed something into her palm. 'Here's a shilling in case you should need it.'
    A shilling! She must be going away for ever! 'Father!' cried Fenella.

The part of this passage I want to draw attention to is 'A shilling! She must be going away for ever!' Now you will note that (as with 'She loved him') these two utterances are ostensibly in the third person; Fenella is described as 'she' and inverted commas are not used. But it is clear that it is Fenella's thoughts that are being given us here as neither the narrator nor her father would be so surprised at her being given a shilling.

Let me return to the issue of the ambiguity associated with the technique, an ambiguity which sometimes makes it impossible definitively to attribute a statement either to narrator or character. Take this extract from Virginia Woolf's *Mrs Dalloway*:

Elizabeth rather wondered whether Miss Kilman could be hungry. It was her way of eating, eating with intensity, then looking, again and again, at a plate of sugared cakes on the table next them; then, when a lady and a child sat down and the child took the cake, could Miss Kilman really mind it? Yes, Miss Kilman did mind it. She had wanted that cake – the pink one. The pleasure of eating was almost the only pure pleasure left her, and then to be baffled even in that!

Whose consciousness is represented in the last three sentences? Is it (i) narrator (ii) narrator (iii) narrator? Or (i) narrator (ii) narrator (iii) Miss Kilman? Or (i) narrator (ii) Miss Kilman (iii) Miss Kilman? It is even possible that these sentences give us Elizabeth's thoughts as she imagines what Miss Kilman is thinking. As with the opening of 'Fanny and Annie', these ambiguities, this indeterminacy is not a flaw on the passage but quite the reverse: we know what Miss Kilman is thinking and we understand the narrator's attitude towards her and her desires without having to attribute a form of words either to the narrator or to the character. Note that a principle of inertia operates in our response to such passages: unless we are given good reason for changing the way we attribute statements to a particular source or consciousness, we tend to go on attributing them to the one already established as the operative one. The technical term for this 'principle of inertia' in narrative theory is *obstination*.

Another technique, one which has something in common with Free Indirect Discourse without having quite the grammatical specificity of this technique, is that of *coloured* discourse or narrative. This involves the 'colouring' of a piece of (normally but not necessarily) third-person narrative with words, phrases, expressions which the reader associates with the verbal habits of a particular character. The technique is closely related to techniques used to mock, mimic or satirize others in speech. The fiction of Dickens contains innumerable examples of it, from which the following (the opening of Chapter 25, 'Book the Second', of *Little Dorrit* [1857–8]) will serve as illustration:

> The dinner-party was at the great Physician's. Bar was there, and in full force. Ferdinand Barnacle was there, and in his most engaging state. Few ways of life were hidden from Physician, and he was oftener in its darkest places than even Bishop. There were brilliant Ladies about London who perfectly doted on him, my dear, as the most charming creature and the most delightful person . . .

The whole of this chapter is characterized by bitter and ironic mockery of the pretensions of those involved, but in this passage we see Dickens actually incorporating a verbal 'signature' from those he is mocking; 'my dear' is, we take it, the sort of frequent hypocrisy to be found in the conversation of the ladies being ridiculed, and its inclusion colours the narrative around it.

6 *Tone and mode*

Before we leave the general topic of narrative technique, two other important terms need to be mentioned: *tone* and *mode*. The term *tone* refers to the attitude of the narrator (and sometimes, by implication, of the author) towards what is narrated. One fails to understand *Gulliver's Travels*, for example, if one does not perceive the satirical tone that pervades the whole work. And, although it is true that a work like Kafka's *The Castle* suggests that life is meaningless and that communication between human beings is impossible, this aspect of the work has to be seen in consort with the tone of deep pity and sympathy for humanity which pervades the novel.

Mode, however, refers to the type of discourse used by the narrator – the sort of issue we considered earlier in connection with discussion of the term 'point of view'. The term is taken from Continental narrative theorists and is sometimes rendered as 'mood', although as 'mood' has other meanings within literary criticism I prefer to use 'mode' in this context. We can sum up what is meant by mood/mode by saying that it refers to the *relationship* between the narrating and what is narrated in a broad sense – both in terms of *what* is perceived and reported, and in terms of *how* this is treated. I will come back to this relationship and to other technical terms in the discussion of plot later on in this chapter.

If these (and other) terms confuse you, don't worry; remember that the important thing is not to learn a complicated set of terms, but rather to become alive to the sort of distinctions the terms point to.

Character

In Chapter 2 I suggested that it was a moot point whether much fiction from before the time of the modern novel could be said to contain characters as we understand the word. For us 'character' is intimately bound up with individualism: a character is unique; not just the property of a person but somehow simultaneously both the person and the sign or token of the person. (See my comments in Chapter 2, p. 20, on the etymological connection with 'character' meaning 'handwriting'.) According to Clemens Lugowski, so long as human existence is dominated by myth (which is timeless and collective) rather than by history, 'there can be no individual characters', for instead of being autonomous individuals 'they are no more than components of a whole'.[12] Myth was, typically, an oral form, whereas the novel, as I have argued in my opening chapter, has an intimate relationship with writing and printing. No literary or artistic form has a closer relationship to what we now know and understand as character, than does the novel.

If we can conceive of a time during which characters as we know them, both in life and in fiction, did not exist, then it will be easier for us to consider a number of other possibilities. First, that how human individuals are represented in books has some connection with how they are conceived by themselves and by others in real life. Second, that as our view of human individuality in the real world changes, so too may our view of fictional character. Third, that just as there are conventions which govern (or at least influence) how speech is represented in the novel (see the end of this chapter), so too there are comparable and changing conventions which govern the depiction of character in the novel. And, finally, that if there was a time in which human beings saw themselves in

[12] Clemens Lugowski, *Form, Individuality and the Novel*, p. 83.

ways untouched by the individualism which forms an important component in the developing concept of fictional character, there might just be
such a time again one day, so that those elements which may strike us as
strange and unnatural in the way in which individuals are represented in
modernist and postmodernist fiction *may* be prophetic of some such
change. (Needless to say, they may not!)

We should certainly not assume that because we can respond so fully
to characters in the fiction of past ages there is, therefore, no change in
fictional characterization from age to age. Not only are there technical
changes in the way novelists learn to create and reveal characters, but
changes in human beings outside literature (or at least the belief that
human beings have changed) often lead novelists to use new methods to
produce a new sort of character.

In her famous essay 'Mr Bennett and Mrs Brown' (1924) Virginia
Woolf states her belief that:

> all novels . . . deal with character, and that it is to express character – not
> to preach doctrines, sing songs, or celebrate the glories of the British
> Empire, that the form of the novel, so clumsy, verbose, and undramatic,
> so rich, elastic, and alive, has been evolved.

But this does not mean that she sets up 'character' as a sort of human
universal, that she claims that the characters which we find in books or
in real life are always the same. Indeed, earlier on in the same essay she
has already made the remark (doubtless part tongue-in-cheek) that 'in or
about December, 1910, human character changed.'[13] Woolf is of course
exaggerating for humorous and other purposes, but it is clear that she is
serious in the main point that she makes in the essay, that changes in
human beings have taken place, and thus changes in the ways novelists
represent human beings must also occur. Thus if we feel tempted to complain (as many *have* complained) that characters in the *nouveau roman*
are less satisfying than those in Dickens's novels, we should pause to ask
whether or not we would be happy with a situation in which the contemporary novel used only the characterization of Dickens. Would the novel
help us to understand contemporary human beings if that were the case?

My earlier discussion of modernism and postmodernism should have
suggested that recent fiction to which the labels 'modernist' or 'postmodernist' can be attached will be likely to portray the human individual

[13] Virginia Woolf, 'Mr Bennett and Mrs Brown'. In *Collected Essays*, vol. 1
(London, Hogarth Press, 1966), pp. 324, 320.

in ways different from those to be found in the classic realist fiction of the previous century. And this is indeed the case. What seems to go first in the characters we find in modernist fiction are individual coherence, self-knowledge, and autonomy. Virginia Woolf's *The Waves* (1931) is a textbook example of this sort of change. Her character Bernard repeatedly utters statements such as 'I am not one and simple, but complex and many'; 'I am more selves than Neville thinks'; 'We exist not only separately but in undifferentiated blobs of matter'. Not surprisingly, Bernard repeatedly asks 'Who am I?', and even describes himself as a man without a self. It needs to be said that *The Waves* does not attempt to universalize Bernard's view of himself: other characters in the novel *do* have a sense of their coherent and autonomous selves. Nonetheless, the novel opens up the possibility that the classic realist view of character is either inadequate or historically limited – perhaps culture-specific. And other modernist novelists have also portrayed character as complex and problematic rather than simple and self-evident. Think of Klamm in Kafka's *The Castle*, who appears to be different to everyone who sees him and each time that he is seen. It seems clear that with Klamm Kafka is concerned to open certain conventional views of human individuality, both as they operate in fiction and as they are to be found in the world at large, to renewed scrutiny.

Postmodernist fiction generally takes this disintegration of character a stage further, playing with the conventions governing the representation of character in works of fiction so as to expose these to the reader's scrutiny. Thus a *cancelled character* (sometimes called an *erased character*) is a character in a postmodernist work who (or which) is introduced in order to make the reader accept him or her as a human individual, but who is later cancelled or erased and revealed to be only the creation of the novelist, a construct who (or, again, which!) has existence only in the pages of the work. Such a refusal to abide by established conventions has to be seen as comparable to the way in which modernist art frustrated realist expectations by perverting those conventions governing the representation of depth in pictures – perspective. The aim is also to make the reader or viewer think about the way in which these conventions govern our understanding both in literature and in the extra-literary world.

These rather extended introductory comments seem to me to be necessary because 'character' may well appear to be one of the least problematic terms with which you have to deal in studying the novel. The proper names we encounter in a novel – Tom Jones, Anna Karenina, Daisy Miller, Huck Finn, Yossarian – seem very much like those proper names with which we meet in everyday life and by which we designate individual human beings.

And yet even if we stop at names we may realize that characters in novels are not quite like real people. In everyday life we sometimes meet a person with an unusually appropriate name: the very tall person called Long or the radio engineer called Sparks. But the peculiar appropriateness of Heathcliff's name, for instance, is surely hardly ever met with in real life. And what about Dickens's Esther Summerson – who acts like a 'summer sun' in *Bleak House*, dispelling the shadows with which the work is, initially, filled? Even 'Tom Jones' with its resolute *lack* of connotations or associations seems extraordinarily appropriate a name for the non-aristocratic, normally healthy hero of Fielding's novel.

'Tom Jones' is a rather different name from 'M'Choakunchild' – the name given by Dickens to one of his characters – however, and this reminds us of an important point: there are different *sorts* of literary character. Of course there are different sorts of people in ordinary life, but it is not this sort of variation that I have in mind. Think of Meursault in Albert Camus's *The Outsider* and Mr Guppy in Charles Dickens's *Bleak House*. Both are young men who have problems communicating with others and who both have an odd relation to their mothers. But there are also important differences between them, and if we were asked to explain these differences we would have to talk not just about them as if they were ordinary human beings, but in a way that recognizes that they are literary characters which exist within very different sorts of novel.

A good test here is to ask to what extent a character from one literary work could be transplanted to another. We have little difficulty in imagining a meeting between, for example, the title character from Jane Austen's *Emma* (1816) and Mr Darcy from her *Pride and Prejudice* (1813, but written 1796/7). This is because the fictional worlds of these two novels are so similar that such a meeting, which is as much a meeting of fictional worlds as it is of characters, does not seem absurd. This is why we have films with titles like *Godzilla Meets King Kong*, but not, outside the realms of parody, with titles such as *Bambi Meets Godzilla*. (I have actually seen a very short film with this title, and of course knew from the title that it *had to be* a parody!)

If we go back to the young men from *Bleak House* and *The Outsider* we realize, I think, that, unless a novelist were concerned to demonstrate something about the incompatibility of different fictional worlds in an experimental or postmodernist work, Mr Guppy could not be transported to the fictional world of *The Outsider*; he could not survive in that novel as we know him in *Bleak House*. My concept of 'fictional world' is not just a matter of setting (about which I talk below). After all, the settings of *Clarissa* and *Robinson Crusoe* are very different, but there is a sense in which their fictional worlds are much less so. The concept refers, in other words, not just to the real-world elements (places, events, history) contained in the works, but also to the representational conventions governing the depiction of these and other things. *Bleak House* and *The Outsider* are, moreover, partly held together by philosophical and ideological elements: views of the nature of the world which enter into the very fabric and texture of the novels. Meursault lives in a world defined by the preoccupations of existentialist philosophy, and Mr Guppy does not. (Note that the important thing is not whether the *individual character* has certain philosophical or other concerns, but whether their *presentation* and fictional existence accord with these concerns.)[14]

Let us try to explore some of the differences that exist between literary characters. We have some well-established terms to draw on initially: major and minor characters, flat and round characters, stock characters, 'types', caricatures, and so on.

We can sum up one important distinction that has a bearing upon all of the above terms: is the writer interested in developing a character so as to *represent something*, or in order to present a *particular individuality*? The distinction is related to that between novels of 'life' and novels of 'pattern' to which I referred much earlier, for it is in novels of pattern that one is likely to find characters who stand for something (a giveaway feature is often the names) and in the novel of life that one finds characters possessed of a distinctive and idiosyncratic individuality. However, even in the latter case we have to face the paradox that a highly individualized character may also stand for something: Dickens's Mr Toots in *Dombey and Son* and Dostoyevsky's Prince Myshkin in *The Idiot* (1869) are 'originals', yet they also stand for certain valuable qualities which contribute importantly to the themes of the two novels mentioned. Moreover, just because a character is a recognizable *type* (another of Dickens's grotesques, one of Dostoyevsky's eccentrics) does

[14] Since writing this, I have discovered Christine Brooke-Rose's novel *Textermination* (1991), in which a conference of characters out of the great works of literature is convened at the San Francisco Hilton, to pray for their continual survival in readers' minds. It is not exactly *Bambi Meets Godzilla*, but it does have Emma Bovary sharing a carriage with Jane Austen's Mr Elton. Of course, the characters' knowledge of themselves as literary characters gives them something in common and something which their originals lack.

not necessarily mean that he or she cannot also be a realized individual personality in the work.

In his book *Aspects of the Novel* (1927) E.M. Forster uses Dickens's Mrs Micawber (from *David Copperfield*) as an example of what he calls a 'flat character':

> The really flat character can be expressed in one sentence such as 'I never will desert Mr Micawber.' There is Mrs Micawber – she says she won't desert Mr Micawber; she doesn't, and there she is.

Mrs Micawber does not change because she is not allowed genuine *interaction* with other people and situations; even though she has as it were dealings with them, she is independent of them.

Let me pause at this stage to observe that we have now dealt with three different ways in which a fictional character can be unlike a real person. First by being defined by the philosophical or ideological underpinnings of their fictional presentation; second by being limited to one essential characteristic; and third by being possessed of a constituting individuality that is unshakeable by events.

The term 'synonymous character' is often applied to recent, especially experimental, novels; synonymous characters are those which are possessed of essentially the same characteristics and functions in a piece of prose fiction, and are thus to all intents interchangeable.

The word 'function' also deserves some attention at this stage. Its literary-critical usage is most associated with the analyses of folk-tales carried out by the Russian theorist Vladimir Propp, who argued that all folk-tales consist of a selection from a finite number of functions, defined as 'an act of a character, defined from the point of view of its significance for the course of the action'.[15] Not all the available functions would be selected by the creator of the tale, but those which were selected had to appear in the tale in a strict order. Such a description seems foreign to those of us more familiar with the novel than with the sort of work with which Propp was concerned, but, just as I argued earlier that although the novel is to be distinguished from the romance nonetheless romance-like elements can be found in the novel, so too it is worth remembering that, although the modern novel is to be distinguished from the folk-tale in terms of – among other things – its individual and individualized characters, nonetheless modern novelists are still capable of using

[15] Vladimir Propp, *Morphology of the Folktale* (trans.) Laurence Scott (Austin, University of Texas Press, revised edn, 1968), p. 21.

characters as a means whereby an 'act significant for the course of the action' can be achieved. In other words, some characters are there to do things; they are primarily *vehicles*, and their significance is limited to what they do.

Thus although *in general terms* we can distinguish the novel from other fictional forms which represent human individuals but are less interested in their particularity and individuality than in what they do or stand for (the type-characters of the fable or parable, for example), we should recall that the novelist does not have to sign a declaration that he or she will never represent human beings unless the primary or exclusive aim is to display and to celebrate their uniqueness and individualism, or alternatively the ways in which they are most like a real, living person. Even those novelists most renowned and admired for their creation of living characters utilize recognizable types: think of Jane Austen and Charles Dickens. And, indeed, because in real life we classify people ('a typical schoolmaster'), we should not automatically assume that depictions of realistic characters can not also be representations of human types, or of characteristics typical of particular groups of human beings.

The critic John Bayley has argued that an author's love for his or her characters 'is a delight in their independent existence as *other people*, an attitude towards them which is analogous towards those we love in real life; and an intense interest in their personalities combined with a sort of detached solitude, a respect for their freedom.'[16] There is a lot of truth in this, and many novelists have written as if their characters were real people who possessed an independence of their creators – doing things that surprised the latter for example. And yet we should take the more extreme claims made for such character independence with a grain of salt, for characters can never be completely independent from the writers to whom they owe their existence. If they start doing things that do not contribute to the novel or short story of which they are a part, then their further existence is jeopardized.

Moreover, those characters which we need to consider primarily as vehicles may not even be there to play a significant part in the action or the plot of a work. In recent years a number of critics have investigated the concept of *projection characters*, that is, characters into whom the novelist projects aspects of him- or herself – often aspects which cannot be acknowledged either to others or to the author him- or herself. (Psychoanalytic theories of projection lie behind the term; the idea being

---

[16] John Bayley, *The Characters of Love* (London, Chatto & Windus, 1960), p. 7.

that what we cannot accept in ourselves we often project on to others.) In Charles Dickens's *Great Expectations*, for example, the character Orlick is like a dark shadow of the hero Pip, and seems almost like a personification of Pip's repressed desires. He stands up to Mrs Joe (who treats Pip so unfeelingly), and eventually strikes her down, and he openly lusts after Biddy, for whom Pip eventually acknowledges his own desire. In a late, climactic scene he announces his intention of murdering Pip himself, thus dramatizing what otherwise would be a purely internal struggle in Pip. If we take into account the considerable extent to which Pip's portrayal (like that of David Copperfield) is modelled by Dickens on his own character and experiences, then we can understand that Orlick can be described as a projection character: whereas the similarities between Dickens and Pip are open and overt, those between Orlick and Dickens are concealed and even denied.[17]

This means that a novelist may use a character for purposes quite other than 'characterization'; to say that there are different types of character is to say in effect that novelists portray human individuals for a range of different purposes. This is why it is a mistake only to talk about characters in a novel as if they were real people; clearly the novelist relies upon our knowledge of and reactions to real people in his or her creation of character, but characters are often created by novelists for purposes other than that of investigating human personality or psychology. They can be used to tell a story, to exemplify a belief, to contribute to a symbolic pattern in a novel, or merely to facilitate a particular plot development. (The same is true of, for example, descriptions of landscape in a novel. These are sometimes written for quite different reasons than for reminding the reader what actual landscapes are like, and in such cases it would clearly be irrelevant to draw attention to the ways in which such a description failed to mesh with 'what real landscapes are like'.)

Furthermore, novelists can create character in a range of different ways. A classic but still useful distinction between two fundamental ways of creating (or 'revealing') character is that which we owe to Percy Lubbock's *The Craft of Fiction* (1921). Lubbock's distinction between 'telling' and 'showing' (a distinction he probably owed to Henry James) is very similar to that which Georg Lukács makes between 'narration' and 'description' in a famous essay entitled 'Narrate or Describe' (1936). (In neither case, it should be said, is the critic talking exclusively about

---

[17] For a comparable account see Roger Sell, 'Projection Characters in *David Copperfield*', *Studia Neophilologica* (LV), 1983.

characterization.) When Jane Austen opens her novel *Emma* in the following way she is *telling* (or in Lukács's terminology describing) rather than showing us what Emma is like:

> Emma Woodhouse, handsome, clever, and rich, with a comfortable home and happy disposition, seemed to unite some of the best blessings of existence; and had lived nearly twenty-one years in the world with very little to distress or vex her.

But when, later on in the novel, we witness the conversation between Emma and Mr Knightley in which he criticizes her rudeness to Miss Bates, we are *shown* the sort of person Emma is through her behaviour and her responses to him. Since Lubbock, critics have generally preferred 'showing' to 'telling' as a method of revealing character, feeling that this method unlocks the life in characters rather than treating them (as Lukács puts it) as inanimate objects. Moreover we feel that we decide what a character is like when we observe him or her behaving in front of us; we can use our critical intelligence and our knowledge of human beings to reach an assessment of them, whereas when we are *told* something we can only take it or leave it. There is no scope for differential responses to the first sentence of *Emma*. (This is why, in real life, we prefer to make up our minds about people through personal acquaintance and observation rather than on the basis of others' assessments.)

If we think of the most memorable literary characters we probably find that we remember them doing or saying things; we do not so much remember being told things about them. Where a narrative comment on a character does stick in our minds it is probably as a result of something other than characterization; irony, moral discrimination, or whatever.

So far as the creation of lifelike characters is concerned, it is very often the interplay between given and explained attributes that makes them fascinating: think of Heathcliff in *Wuthering Heights*; we understand partly why he is as he is because we have witnessed key elements of his upbringing. But there are some elements that seem to be beyond such an explanation.

I repeat, however, that creating a lifelike character may not always be what a novelist is aiming at – even in pre-modernist or pre-postmodernist fiction. It is not lifelike for Gulliver to be so acute and humane at one part of *Gulliver's Travels* and so obtuse and unconcerned about talk of human suffering at others. But it seems apparent that Swift's primary aim here is not the creation of a consistent and lifelike character so much as the creation of a character whose alternating qualities and abilities will represent different aspects of that 'human race' that is perhaps the real subject of *Gulliver's Travels,* and the creation of a mechanism to display those aspects of the humanity which he is anxious to expose to satire.

What are the most important methods of characterization available to the novelist? I would suggest four that are worth thinking about. First is *description* or *report.* In Conrad's *Heart of Darkness* we know a very large amount about Mr Kurtz before ever he appears before us; other characters in the novella have talked about him so much, have reported on his actions and beliefs, that we feel it is almost as if we had met him ourselves. The description of physical characteristics – and especially of physiognomy – is a very traditional means whereby the writer can suggest what sort of character with which we are faced. Here is a description taken from the second paragraph of Philip Larkin's novel *Jill* (1946). The novel's main male character, John Kemp, is travelling by train to Oxford:

> It looked cold and deserted. The windows of the carriages were bluish with the swirls of the cleaner's leather still showing on the glass, and he confined his eyes to the compartment. It was a third-class carriage, and the crimson seats smelt of dust and engines and tobacco, but the air was warm. Pictures of Dartmouth Castle and Portmadoc looked at him from the opposite wall. He was an undersized boy, eighteen years old, with a pale face and soft pale hair brushed childishly from left to right. Lying back against the seat, he stretched his legs out and pushed his hands to the bottom of the pockets of his cheap blue overcoat. The lapels of it curled outwards and creases dragged from the buttons. His face was thin, and perhaps strained; the expression round his mouth was ready to become taut, and a small frown lingered on his forehead. His whole appearance lacked luxuriance. Only his silky hair, as soft as seeding thistle, gave him an air of beauty.

Just imagine that you have met someone on a train, and when discussing this meeting with someone you know, later on, you start to describe him. You would, I venture to suggest, hardly come out with something along the lines of the above paragraph. If you did it would probably be because you were a student of literature indulging in a little parodic behaviour. To put the matter another way, we can say that Larkin's description of John Kemp depends upon certain fictional conventions which regulate the depiction of character, conventions which require that physical characteristics be seen to have a much more reliable function as indicators of non-physical characteristics (and even, perhaps, of past or future events and experiences) than they do in real life. (Which is not to deny that the *sort* of observation we are presented with in this passage, and the *sort* of deduction implied by these observations, are closely related to those observations and deductions we make in real life. Railway journeys seem to encourage this sort of speculation about other people: Virginia Woolf makes this the basis of her essay 'Mr Bennett and Mrs Brown'.)

Thus from this passage we can detect that John Kemp is not rich, and is not a thrusting, self-confident youth ready and eager for heroic exploits. We assume that the journey he is making has not been made by him before, otherwise he would not notice so many details about his surroundings. (We are not told that it is he who sees and smells these details, but we are encouraged to assume this.) We deduce too that behind his lack of position and self-confidence there reside qualities which now are apparent only to the discriminating and which will allow him to deal well (finally, perhaps) with whatever trials he has to face. It is almost unthinkable that this is a description of a man who will turn into a pompous and conforming hypocrite, or who the reader will end up hating. Behind his portrayal we may, even at this stage, recognize what can be called a mythic element: the unregarded person who has a potentiality that, when revealed, may surprise those who take him at face value. That silky hair of John Kemp's, along with the expression round the mouth and the frown on his forehead, betoken much more on the first page of a novel than they would in real life: they stand for that unregarded quality or virtue which exists only *in potentia*, whose emergence the novel will trace. As the description also insists on the shabby cheapness of Kemp's clothes and of the third-class compartment we can feel reasonably confident, too, that the novel will pay particular attention to the problems raised by his background and social class – especially as the first paragraph of the novel has informed us that the train he is on is heading for Oxford. Notice how much we know of Kemp and how much we are already expecting – and he has said nothing and done nothing so far!

Second, character can be established by *action*; when Insarov in Turgenev's *On the Eve* (1859) throws the insolent German into the water – an action of which his effete Russian companions are palpably incapable – then we learn something about him which pages of description could not give us.

Third, through a character's *thought* or *conversation*. Dialogue in particular is a wonderful way of revealing character: think how much we learn about Miss Bates in *Emma* merely through her conversation – so much so that comment from Austen's narrator is really not needed. Modern novelists have shown how much we can learn about a character merely by following his or her thoughts; in Virginia Woolf's *Mrs*

Dalloway Clarissa Dalloway and Peter Walsh actually *do* very little, but by the end of the novel we feel that we know them quite well just by having followed so many of their thoughts.

In the third paragraph of Chapter 2 of Jane Austen's *Sense and Sensibility* (1811) we learn as much as we feel we need to learn about Mrs John Dashwood through a report of her comments on her husband's intended generosity to his two sisters. Austen here uses that variant of Free Indirect Discourse known as represented speech (see p. 109), and as this is frequently used for ironic purposes we are alert to the hints concerning the narrator's disapproval without their ever becoming overt and intruding directly. The paragraph reads:

> Mrs John Dashwood did not at all approve of what her husband intended to do for his sisters. To take three thousand pounds from the fortune of their dear little boy, would be impoverishing him to the most dreadful degree. She begged him to think again on the subject. How could he answer it to himself to rob his child, and his only child too, of so large a sum? And what possible claim could the Miss Dashwoods, who were related to him only by half blood, which she considered as no relationship at all, have on his generosity to so large an amount. It was very well known that no affection was ever supposed to exist between the children of any man by different marriages; and why was he to ruin himself, and their poor little Harry, by giving away all his money to his half sisters?

In addition to represented speech, the passage contains a number of textbook examples of *colouring* (see p. 112); although the narrator is telling us about Mrs John Dashwood, the passage is full of what the reader recognizes as the character's own speech characteristics. We may observe how close the technique is to mimicry in spoken delivery.

The pleasure we get from such a passage is related to the fact that, while on the one hand we do not feel we are being told how to regard Mrs John Dashwood and coerced into a particular attitude towards her, we also feel that our amused disapproval of the character is shared by Jane Austen (through her narrative voice). The very omission of comment on the more outrageous statements of Mrs John Dashwood's indicates a narrative opinion that they *are* so outrageous as to require no comment.

Finally the novelist can use *symbol* or *image* to reveal and develop a character. In Jean Rhys's novel *After Leaving Mr Mackenzie* (1930) the heroine, Julia, has just been to see a Mr James, shortly after visiting her dying mother and her spinsterish sister Norah:

She wanted to cry as he went down the stairs with her. She thought: 'That wasn't what I wanted.' She had hoped that he would say something or look something that would make her feel less lonely.

There was a vase of flame-coloured tulips in the hall – surely the most graceful of flowers. Some thrust their heads forward like snakes, and some were very erect, stiff, virginal, rather prim. Some were dying, with curved grace in their death.

The tulips present us symbolically with the different options we have seen women in the novel plump for: snake-like cunning and self-interested behaviour; prim and virginal lifelessness (like Norah); and death – like Julia's mother. Such a passage contributes to our knowledge of and attitudes to characters in the novel even though nothing is said directly about any character.

Plot

Let us start with a definition: a plot is an ordered, organized sequence of events and actions. Plots in this sense are found in novels rather than in ordinary life; life has stories, but novels have plots and stories. As E.M. Forster puts it, a story is a narrative of events arranged in their time-sequence, whereas a plot is a narrative of events with the emphasis falling on *causality*. Not all commentators would agree that causality is the distinguishing feature, but all would agree that there is a necessary distinction to be made between the incidents about which we are told in a novel in their chronological order, and the actual narrating of these events in perhaps quite a different order in the novel. I prefer the terms 'story' and 'plot' to describe these two, but some critics use terms created by a group of theorists known as the Russian Formalists, the terms 'sjužet' (plot) and 'fabula' (story).

Recent narrative theorists have performed a useful function in providing terms and concepts such that we can more accurately analyse how and why the plot of a novel deviates from strict and simple chronological progression.

First of all are a number of terms connected to what is termed *order*. The plot of a novel may move backwards and forwards in time, instead of proceeding steadily forward in chronological order. Any deviation from such strict chronological progression is termed an *anachrony* by narrative theorists, and there are a large number of such deviations possible. The most frequent are *analepsis* (or flashback) and *prolepsis* (or flashforward). The italicized terms are perhaps to be preferred, as 'flash' suggests a rather short movement backwards or forwards in time whereas an analepsis or a prolepsis may be of very considerable duration. These terms are also utilized by some theorists when a past or future event is evoked rather than actually invoked or brought about. Thus we can say that, at the start of Joseph Conrad's *Nostromo* (1904), the fate of the *gringos* who die searching for silver is proleptic of much that follows in the novel, even though it does not technically involve events that take place after the time established at the novel's opening.

Second, a novel's plot may include gaps, omissions, absences. These can be referred to collectively as *ellipses* – the same term as is used to refer to the succession of dots in a text that indicates that something has been omitted. Thus in *Wuthering Heights* we never get to know what Heathcliff does after his sudden disappearance and up to the time of his reappearance. This is what can be termed a (relatively) *unmarked* ellipsis – in other words, the text does not display the fact that something is not there. But when in Charles Dickens's *Bleak House* Esther Summerson seeks to explain why she finds Mrs Woodcourt irksome, and breaks off with the words, 'I don't know what it was. Or at least if I do, now, I thought I did not then. Or at least – but it don't matter', then we have a clearly *marked* (or explicit) ellipsis: the reader's attention is drawn to the fact that something that is known to the narrator is withheld from him or her. A novelist typically uses a marked ellipsis to get the reader's imagination working: what has happened here? Why are we not told? An unmarked ellipsis usually involves the novelist's having skipped over a period of time during which nothing of artistic significance can be represented as having happened.

Third, the element of *duration* is also of great significance. Nearly all works of prose fiction vary the relationship between narrating and narrated time. (The term *tense* is sometimes used to refer to this relationship, although this usage – influenced by structuralist theorists – has not achieved general acceptance in Anglo-American circles.)

Whatever term we use, the essential characteristic involved is crucial to how novels work. A novelist can spend fifty pages to tell us about one day in the life of a heroine and then ten further pages to cover fifty years. Such extremes are sometimes referred by means of the linked terms *scene* (a passage in which there is a sort of equivalence between narrated and narrating time – for example a dialogue in which all the words exchanged are given) and *summary* (where it is clear that the narrative is reducing all the potentially available material to a descriptive outline). Within a single novel the ratio of narrated to narrating time can alter very substantially, and it is worth paying careful attention to such shifts.

In *Tristram Shandy*, for example, Lawrence Sterne makes frequent jokes about the relationship between narrated and narrating time: at the start of Chapter 21 for example the text breaks off into a digression in the middle of a sentence uttered by uncle Toby, to utter which he has had to take his pipe out of his mouth. The digression finished, the text continues: 'But I forget my uncle Toby, whom all this while we have left knocking the ashes out of his tobacco pipe.' The suggestion that narrated and narrating time are coterminous is a joke, we know, because in many novels the narrator can leap over enormous gaps of time. Think how few pages are devoted to the period of Cathy's married life prior to the return of Heathcliff in *Wuthering Heights*.

Fourth, narrative theorists also isolate the important topic of *frequency*, and note the following fundamental possibilities:

(i) one event narrated once (*singulative frequency*)
(ii) a repeated event narrated the same number of times that it occurs (*multiple frequency*)
(iii) one event narrated many times (*repetitive frequency*)
(iv) many events narrated once (*iterative frequency*)

(i) is probably the most usual; many events that are depicted in a novel are recounted to the reader only once. But the same event can be recounted many times, as in (iii). Sometimes this has the effect of sharpening our understanding: the detective novel typically recounts the same events more than once, so that what was unclear the first time becomes clear in the detective's final version. In *Catch 22*, as Snowden's death in the bombing raid is as it were 'replayed' repetitively through the anti-hero Yossarian's consciousness while the novel unfolds, the reader is reminded of a psychoanalytic patient recalling more and more of a repressed event. And, as the reader gradually comes to learn what happened, the repetition conveys the traumatic effect that the event has had on Yossarian. But sometimes clarity is not the intended result: the repeated but different depictions of the same events in Robbe-Grillet's *The Voyeur* (1955) lead the reader to lose confidence in the existence of a single, isolable series of events.

Many similar events may, as in (iv), be recounted as one. In Conrad's *Lord Jim* Marlow's narrative is introduced as an *iterative* narrative – in other words, we are told that what we are about to read is an account that has been given many times. But by the time the account ends it has become a single, particular act of recounting. In the same novelist's *Nostromo*, the shifts between the singulative and iterative modes are technically contradictory, but they draw attention to important thematic concerns about the uniqueness of events and the repetitions of history. It seems clear that in this novel Conrad is deliberately and repetitively frustrating his readers' expectations: passages change unexpectedly from singulative to iterative mode, and from iterative to singulative mode.[18]

The French narrative theorist Gérard Genette has isolated a variant on the above alternatives which he calls the 'pseudo-iterative'. According to his definition, the pseudo-iterative occurs when 'scenes presented, particularly by their wording in the imperfect, as iterative', are possessed of a 'richness and precision of detail [which] ensure that no reader can seriously believe they occur and reoccur in that manner, several times, without any variation'.[19] The following passage from Guy de Maupassant's novel *Bel Ami* provides a representative example of this technique. (In its original French this passage also contains examples of the pseudo-iterative, although perhaps because of the problems this mode presents to the translator, frequency in the English version does not match frequency in the French original exactly.)

> So they would go into low dives and sit down at the end of the squalid murky room on rickety chairs at decrepit wooden tables. An acrid cloud of smoke and the smell of fried fish filled the room; men in their working clothes were shouting at each other as they downed their glasses of raw spirits; and the waiter would stare at the strange couple as he put the two cherries-in-brandy down in front of them.
>
> Scared and trembling, but blissfully happy, she would begin sipping the red fruity liquid, looking around her, bright-eyed but uneasy: each time she swallowed a cherry she felt she was doing something wrong, every drop of the spicy, burning liquid that slipped down her throat gave her a sharp pleasure, the pleasure of forbidden fruit, of being naughty.
>
> Then she would whisper: 'Let's go.' And so they would leave. She slipped away hastily, tripping along and holding her head down, like an actress leaving the stage, between the drinkers sitting with their elbows on the table, who looked up with a suspicious, surly air as she went by. And when she was through the door, she gave a big gasp as if she had just escaped from some terrible danger.

---

[18] See my article 'Repetitions and Revolutions: Joseph Conrad's Use of the Pseudo-iterative in *Nostromo*', in the special Conrad issue of *La Revue des Lettres Modernes* (Paris, Editions Minard, 1997).

[19] Gérard Genette, *Narrative Discourse: An Essay in Method*, (trans.) Jane E. Lewin (Ithaca, NY, Cornell University Press, 1980), p. 121.

Sometimes, with a shudder, she asked Duroy:
'What would you do if they insulted me in a place like that?'
He said airily:
'Defend you, of course!'[20]

It seems clear that the detailed dialogue provided here is not fully reconcilable with the suggestion that it represents what was said again and again on different occasions. We may repeat ourselves, but not normally with this degree of accuracy. Like many novelists, Maupassant is I think concerned to have the best of both worlds here. He wants to give the reader a sense of repeated – even obsessively repeated – actions, while giving them enough individualized flavour to dramatize them and to prevent the reader from getting bored. What we have in this passage, then, is the dramatic sense of a *particular* scene, and the thematic insistence upon repetition.

When someone sees that we are reading a novel and asks 'what happens?', we generally simplify its plot so as to bring it back more to resemble the story behind the plot. It is often only when we do this that we recognize that novels with very different plots may have rather similar stories. Thus if you have read William Faulkner's novel *Absalom, Absalom!* (1936) and Jean Rhys's novel *Wide Sargasso Sea* it may have struck you that the stories in both novels are remarkably similar (which you can confirm if you wish by writing them both down in chronological order). But the two plots are very different, so different that the similarity of the stories is not immediately apparent.[21]

Again, if you have read Joseph Conrad's novel *Nostromo* then you will know that it has a very complex plot; the novel *could* have been told in a far more straightforward chronological way. So why did Conrad tell it in such a complicated, *unchronological* manner? Come to that, why did Emily Brontë not have *Wuthering Heights* told to us in one consistent chronological sweep – perhaps by having Nelly Dean recount it all at one meeting to Mr Lockwood?

The answer to such questions has to be complex. A novelist constructs a plot in a particular manner so as to draw attention to certain

---

[20] Guy de Maupassant, *Bel Ami*, (trans.) Douglas Parmée (Harmondsworth, Penguin, 1975), pp. 124–5.

[21] In my *Multiple Personality and the Disintegration of Literary Character* (London, Edward Arnold, 1983), p. 91, I refer to similarities between the *plots* of these two novels whereas, in the light of the point I am making here, I should strictly speaking have talked of similarities between their *stories*.

things which might otherwise escape notice, to produce a different effect upon the reader, and so on. One result of the complex plot of *Nostromo*, for example, is that Conrad deliberately kills what many novelists strive for: tension and reader expectation. We know from early on in the novel what the eventual outcome of the political and military struggles in Costaguana will be. Why does he do this? One likely reason is that as the reader has to devote less attention or expectation to *what* will happen, he or she is able to devote much more to *why* and *how* it happens. Tension may make us read quickly and carelessly, anxious only to find out what happens. (We rarely read the closing chapters of a detective novel with the attention that the opening chapters command.) Such a reading is actively discouraged by the way *Nostromo* is constructed. It is encouraged, in contrast, by the way in which most of Dickens's novels are constructed; Dickens wants to engage our sympathetic anxiety and so he sets puzzles, creates mysteries which we read on in order to solve.

If a novelist abandons strict chronology, how else can he or she retain coherence, make the novel hang together? Well, as Forster points out, *causality* is one important possible way. We are still interested as we read the final chapters of *Nostromo* because we want to see why things happen as they do; causal links are revealed to us.

Apart from causality, a novelist can draw *parallels* and *resemblances* between characters, situations and events such that the novel has coherence even if it plots neither chronological sequences nor causal relationships. It is only to a limited degree that we believe that the experiences of the second generation characters in *Wuthering Heights* are actually caused by the lives and actions of the first-generation characters. But the parallels (with variations) and resemblances between the younger Cathy, Linton Heathcliff, and Hareton Earnshaw and their parents make us feel that this latter part of the novel continues themes and enquiries from the first part of the novel. Moreover, it is important for the reader first to encounter Heathcliff at the stage at which he is encountered; this means that we see the young Heathcliff in the context of what we know he will later become, and thus our attention is focused in very particular ways as Nelly Dean starts to tell Lockwood of the childhood and young adulthood of Heathcliff and Catherine.

Alternatively, a novel can be held together by a common character or event. The picaresque novel and the *bildungsroman*, for instance, are held together by the personality of the central figure. The extent to which this provides a complex or a crude unity depends to a certain degree upon the complexity of the central character.

It should be said that with modernism and postmodernism the whole question as to whether a novel *should* compose a unity has been raised. Virginia Woolf argued against plot (in the sense of the traditional, well-structured, realist plot) as one of a number of distorting conventions in her essay 'Modern Fiction' (1919), and a novel such as Alain Robbe-Grillet's *The Voyeur* certainly lacks unifying elements as we have outlined them. The actual denial of causality as a reliable principle by many modernist and postmodernist novelists has, inevitably, affected the way they construct their novels.

We can describe plots in two ways: either in terms of the dominant human activities which form the motivating principle in them or which are induced in the reader by them, or in more technical ways. In the first category we can include plots structured around *conflict* as in many ways the plot of *Nostromo* is; around *mystery* as are many of Dickens's novels; around *pursuit* or *search* as is *The Castle*; around a *journey* as is *Gulliver's Travels*; or, finally, around a *test* as is Joseph Conrad's *The Shadow-Line*. Now of course these are very simplistic descriptions and we would want to say that all of these novels are structured around far more complex issues than the single topics mentioned. But it is worthwhile remembering that a novel is often given force and coherence by a dominating element such as one of these plot types provides. *The Shadow-Line* is much more than just a test, but the theme of the test is a sort of archetypal bedrock within the organizing logic of the plot, and we should not be ashamed to see it as such.

A more technical classification of plots will provide us with terms like 'picaresque/episodic'; 'well-made' (the traditional nineteenth-century realist plot)'; 'multiple' (many novels have two or more lines of plot, sometimes interconnecting and sometimes not. It is very often important to be able to single out a *main plot* from its attendant *subplot[s]* for the purpose of analysis.)

You may recall that much earlier on in this book I quoted Johnson's remark that if one were to read Richardson 'for the story' then one would be so fretted that one would hang oneself. Plot and story fulfil different sorts of functions in different novels; according to Johnson the story in a novel by Richardson is there only to give occasion to the sentiment; in other novels the tension and suspense relative to our desire to know 'what happens next' are an integral part of the appeal of the work. When the 'outer' narrator of *Heart of Darkness* warns us (indirectly) that we are about to hear 'about one of Marlow's inconclusive experiences' then we need to adjust our expectations with regard to the plot of the novella. In contrast, modernist fiction typically has inconclusive endings, endings which leave the reader perhaps puzzled and unsatisfied, but puzzled and unsatisfied in ways that are productive of further thought. You should, again, remember that such matters go beyond the merely technical. Nóvels without omniscient narrators and with inconclusive endings are perhaps less likely to betoken a conventionally Christian world-view in which everything is concluded in the manner in which God wishes; a novelist whose novels all have happy endings is unlikely to suffer from existential doubt.

Structure

Structure and plot are closely related to each other, and it might have made sense to include this section as a subsection of 'Plot'. But the term 'structure' does, properly, refer to something rather different from plot. If we can think of the plot of a novel as the way in which its story is arranged, its structure involves more than its story, encompassing the work's total organization as a piece of literature, a work of art. Nor are the terms 'structure' and 'form' to be confused; the latter term does not normally include thematic elements in the work (see my comments later on concerning 'theme') whereas these are involved in a novel's structure. Structure involves plot, thematics, and form: it refers to our sense of a novel's overall organization and patterning, the way in which its component parts fit together to produce a totality, a satisfying whole – or, of course, the way in which they fail so to do.

Let us start by observing that different novels have very different sorts of structure. We feel, for example, that some parts of *Moll Flanders* or *Huckleberry Finn* could be shifted around in position without making too great a difference to the works in question: how many readers of the former work remember clearly whether Moll robbed the little girl of her necklace before or after she robbed the drunken man in the coach? But to shuffle around the parts of *Wuthering Heights*, or of Henry James's *What Maisie Knew* (1897) would, we feel, do something rather serious to the works in question. Clearly the fact that a novel has an *episodic* structure has an important bearing upon such issues; if a work is structured around a series of relatively self-contained episodes, then these can be assembled in different orders without making too much difference – just as on a modular degree scheme with self-contained modules one can often take different courses in any order one chooses.

But the matter goes beyond plot, as I have suggested. In *Moll Flanders* there is very little alteration in Moll's character, or in the values the narrative underwrites, or in the symbolic meanings contained in the work (if there are any worth noting in this novel!). In short, *Moll Flanders* is consistently structured on the principle of 'repetition-with-slight-variation'. There is somewhat more development and change in *Huckleberry Finn* – in the characters and relationship of Huck and Jim, in the general *tone* of the novel subsequent upon the experiences Huck has, and so on. Thus although some of the scenes in this novel might be switched around without too much effect, such alteration would have to be more limited than with *Moll Flanders* if we wished to avoid damaging the work. But the intricate patterning of a novel such as *What Maisie Knew* would, surely, be completely destroyed if we started to move sections of it from place to place.

Very often the *chapter* and *section* divisions made by the author impose a structure upon a work – or bring out one that is implicit but not overt in it already. It is interesting to read Conrad's *The Shadow-Line* in his manuscript version, in which there are no section divisions, and then to see how differently the published text of the novel reads with these divisions included. Very often such divisions perform the useful function of telling the reader when he or she can pause and put the book down for a bit, and, as it is at these points of time that we think backwards over what we have read and forwards to what we hope for or expect, such divisions can be very significant. It is doubtless for such reasons that Virginia Woolf disapproved of the 'ill-fitting vestments' of the 'two and thirty chapters' of what she called the materialist novel (in her essay 'Modern Fiction'). If clothes or vestments 'make the man', then the wrong chapter divisions may make the novel something other than its author wants.

Order and *chronology* – issues upon which we touched when talking of plot – can be crucial to the matter of structure. The difference between a novel's 'story' and its 'plot' can tell us much about its structure. It is often an interesting exercise to map out a novel's story and plot in note form one above the other.

But structure as I said involves thematic elements too. Note how the repetition of thematic elements in Charles Dickens's *Bleak House* helps to structure that work. Just to take one example: Esther Summerson, like her mother Lady Dedlock, has to choose between two men: a rich, older, 'safe' one and a younger, less wealthy, more 'risky' match. But, whereas Lady Dedlock makes what we are led to see as the wrong choice in marrying the older man, Esther – after an initial false decision – makes what is clearly the right decision for her. Now this produces an element of *pattern* in the work which contributes to our sense of its structure, a pattern which blends in with other things in *Bleak House* to produce a satisfying work of art.[22]

Take a different example. In Katherine Mansfield's short story 'The Voyage' we have a very simple story: a little girl leaves her father at a New Zealand port and travels by boat with her grandmother to the other island of New Zealand, to her grandparents' home. In the course of the voyage we discover that her mother has died, and the reader assumes what Fenella – the girl – has not yet realized: that henceforth she is to be brought up by her grandparents. What is striking about the story is that whereas images of darkness and cold dominate the opening of the story, these gradually give way to images of light and warmth which (especially the images of light) dominate the close of the story. Now clearly this shift of images contributes to the structure of the story: without actually being told it directly we realize that Fenella is moving out of an unhappy period of her life into a potentially much happier one: structurally the voyage is not just from one *place* to another, but this is complemented by travel from one *state* to another.

Structure involves ideas and sensations of some sort of pattern: *completion, reiteration, contrast, repetition, complementarity* – all of these and others can be invoked in us by a work's structure. And we should note that the *frame* of a narrative – what constitutes its outer limits – can contribute importantly to our sense of structure. Consider the narrative framing of the 'tale within a tale' of *Heart of Darkness* and *The Turn of the Screw*. Think of the importance of the fact that the events recounted in James Joyce's *Ulysses* and Virginia Woolf's *Mrs Dalloway* take place within twenty-four hours. Note the thematic framing effect of our being told in E.M. Forster's novel *Howards End* that the work is not to be concerned with the very poor, who are 'unthinkable'. In each case our view of what is 'in' the novel is given structure by our sense of what is excluded from it.

Setting

'Setting' is one of those terms about which recent literary critics have felt increasingly uneasy. Doesn't the term suggest a perhaps too-simple relationship between characters and action on the one hand and the context within which these take place on the other? Doesn't it sound

---

[22] A very useful study of the rôle of repetition in fiction is J. Hillis Miller's *Fiction and Repetition: Seven English Novels* (Oxford, Blackwell, 1982).

rather unsatisfactory to talk about the Nottinghamshire 'setting' of D.H. Lawrence's *Sons and Lovers* or the Yorkshire 'setting' of *Wuthering Heights*, as if the same actions might conceivably have taken place elsewhere – in Tunbridge Wells or Minnesota? The fact that so many characters in Emily Brontë's novel have names that are also the place-names of towns and villages around her native Haworth suggests a relationship between character and environment too organic, we feel, to be described with the term 'setting'.

There may well be other reasons why this has become a relatively unfashionable term of late. Much twentieth-century fiction, especially that which we can term modernist or postmodernist, presents a view of human beings as symbolically homeless, deracinated, alienated from their environment. That close bond between individual and place that is celebrated in the regional novel is harder to find in our own time than it was in the nineteenth century. As a result, many novelists present the reader with settings whose unfamiliarity and unwelcoming aspect is frequently generalized. If all large towns are unwelcoming and dehumanizing, then does it matter which large town one is in? It is surely an essential part of the force of Franz Kafka's *The Trial* (1925) that the setting is not specified in real-world or real-time terms. The point is that in our age the events depicted could take place in many different locations: our time is the time of universal human experiences.

And yet even today it is important to be aware of the context within which the action of a novel takes place – and this does not just mean its geographical setting; social and historical factors are also important. It is just as important to ask why the author has chosen the setting he has chosen when it is generalized (as in the case of *The Trial*) as when it is highly specific (as with, say, Walter Greenwood's *Love on the Dole* [1933]).

To start with we need to distinguish between realistic and conventional or stylized settings. The famous country-house of the classic detective story is obviously a highly *conventional* setting; we are not interested in the particularity of such country houses and their environments, they serve, rather, the function of providing a stylized and familiar setting within which a conventional set of happenings can unfold. Many detective stories actually choose relatively artificial settings that have the function of isolating the characters (or suspects) from the outside world: the closed world of Agatha Christie's Orient Express stuck in the snow is a paradigm case.

At the other extreme we can cite highly realistic settings like that of the tuberculosis sanatorium that dominates Thomas Mann's *The Magic*

Mountain. Here, however, we need to tread carefully, for although this may be a realistic setting it is a very *symbolic* one as well. It is not hard to see the sanatorium full of sick people as representative of pre-First-World-War Europe with its sicknesses and fatal illnesses. Authors are very often quite conscious of such symbolic meanings; in his essay 'Well Done' (1918) Joseph Conrad refers to 'the ship' as 'the moral symbol of our life', and clearly we need to take such a statement into account when looking at those of his works which are set on board ship.

Sometimes the choice of a suitable setting helps an author to avoid the need to write about things that he or she is not good at, or interested in, writing about. It was convenient for Conrad, for instance, that his ships often contained no women. A setting in the historical past can often help an author to avoid contemporary issues about which he or she feels confused; the setting that E.M. Forster chooses for *Howards End* enabled him to avoid writing about the very poor. It is generally agreed that Jane Austen chose settings for her novels which allowed her to exercise her strengths and conceal her weaknesses so far as her knowledge of different sorts of people and of human experiences was concerned.

Moreover, Dickens's frequent choice of London as setting for his novels was convenient in other ways: the mass of concealed relationships, indirect forms of human communication, and innumerable secrets to be found in London offered a perfect opportunity to a novelist whose plots contain all of these elements in abundance. Dickens understood that human values and experiences may be displayed in the physical environment – both in real life and in fiction. A novel such as *Bleak House* suggests that the physical state of the London streets mirrors and displays the values and inner lives of the people and institutions to be found on and around them. A classic example of this vision can be found in this novel's opening pages, pages that have been analysed so often that further discussion of them is probably not necessary.

Just as there are conventions relating to the description of individual human appearance (see the previous section on character), so too descriptions of places and environments are conventionally seen to denote something about the people associated with them. Think of the descriptions of Wuthering Heights and Thrushcross Grange in *Wuthering Heights.*

A setting can also be a crucial factor in the creation of *mood* or *moral environment.* (Note that *mood* here is being used in its ordinary sense, and not in the technical sense outlined above in the discussion of narrative technique. To avoid possible confusion I prefer to reserve the term *mode* for discussions involving narrative technique.)

If we think of *The Great Gatsby* we can see, I think, how a setting can make an essential contribution to a work's mood. And this example reminds us that theme and subject and setting can be inextricably intertwined: you could no more set *The Great Gatsby* in the northern England of the 1930s than you could set *Love on the Dole* anywhere else.

Remember that there is a difference, in this context, between 'mood' and 'tone': the latter term involves narrative *attitudes towards* what is recounted and described. A setting may help to create a particular *mood* in a story, but only the narrative treatment can confirm a certain *tone*.

Theme

'Theme' is a much-used word in the literary criticism of the novel, and a favourite word for use by lecturers and teachers in essay and examination questions. 'Discuss the treatment of the theme of evil in *Crime and Punishment*'; 'Write about the theme of escape in *Huckleberry Finn*', 'the theme of alienation in *The Castle*', and so on.

Students often find such questions or topics baffling. What exactly is a theme? Well, the confusing answer to this question is that the term is used in a number of different ways.

First, some critics find it useful to distinguish between theme and thesis. The simple distinction here is that although both pose questions, a thesis also suggests or argues for answers. A theme, in contrast, can involve the establishing of a set of issues, problems, or questions without any attempt to provide a rationale or answer to satisfy the demands these make of the reader. Traditionally, novels dominated by a thesis have been valued less highly than those in which certain themes are raised or treated: in contrast to earlier generations of readers perhaps, some recent critics have preferred our novels not to be overtly didactic, to be open-ended rather than pointed towards solutions at which the author has already arrived. We should ask whether such an attitude is always justified; novels that are fired with their creators' crusading zeal or commitment to a belief or a cause constitute a very substantial part of the body of fiction, and since its birth the modern novel has had a significant commitment to didacticism.

I should say that the problem with making too simple a distinction between theme and thesis is that it is not universally accepted: for some commentators 'theme' is something of an umbrella concept and includes what we have just termed thesis. In common with a number of other terms, then, 'theme' can be defined in a broad, more inclusive and weaker sense, and a strong, narrow and exclusive one.

If we define theme in the former, weaker sense then we will not be surprised to discover that a large and complex novel can have a range of varied themes attributed to it. Charles Dickens's *Bleak House*, for instance, has been variously interpreted as containing the themes of 'parental responsibility', 'the heartlessness of the law', 'the evil of "causes"', 'the destructiveness of choosing money and position rather than love', 'the centrality of writing to Victorian society' – and many more. It needs to be remembered that a complex novel is likely to be susceptible of analysis in terms of a large number of different – perhaps interlocking – themes. All of the aforementioned themes *can* be found in *Bleak House*, but the reader must decide what their relative force and importance in the novel are.

Second, we should note that a theme may be overt or covert. That is to say it can be either consciously intended and indicated as such by the author, or alternatively, discovered by the reader/critic as an element in the novel of which perhaps even the author was unaware. (If we retain the distinction between theme and thesis then it will be understood that a thesis is much more likely to be consciously intended than a theme.) Thus although we can be pretty certain that Saul Bellow had the *carpe diem* theme ('Live today, while you can') consciously in mind in his novella *Seize the Day* (1956) – because the title makes this much clear – we cannot be so sure that Alan Sillitoe had the theme of working-class socialization equally in mind with regard to the writing of his *Saturday Night and Sunday Morning*. (And we should also bear in mind the possibility that Bellow recognized his theme and chose his title after he had completed writing his story.)

So far as Sillitoe's novel is concerned, a sociologist could certainly find in it the process whereby a rebellious working-class young man is led in representative stages to accept those norms of behaviour, attitudes and institutions which most young working-class men in Britain in the late 1940s and 1950s were led to accept. But as to whether Sillitoe consciously intends this, and whether it is an overt theme of the novel, is a matter for debate. (Given the novel's origin in a number of discrete pieces it seems rather unlikely.)

Symbol and image

Let us introduce this section with some concrete examples. In E.M. Forster's *Howards End* the motor-car plays an important rôle. We could respond to this fact by pointing out that the car had not been around for very long at the time that the novel was written, and that Forster was merely incorporating a piece of contemporary reality into his novel for the purpose of increased verisimilitude. Few readers of the work would find this satisfactory as an explanation. The motor-car in *Howards End* clearly *stands for* or *represents* something; it is not merely a means of transport but a *symbol* in the novel. By this we mean that it carries with it various ideas, associations, forms of significance that in ordinary life it might not have in people's minds: 'the new and destructive of the traditional'; 'the mechanical as against the organic'; 'unfeeling social change'; 'violence and death'; 'the selfish pursuit of personal comfort by the rich' – and so on.

Notice that I have not suggested that the car in *Howards End* stands for just one, fixed thing; it is characteristic of symbols that they do not have a simple one-to-one relationship with what they stand for or suggest. (In this respect a symbol can clearly be distinguished from an allegory, the secondary meaning of which is usually single and distinct. Thus the allegorical meaning of Bunyan's *Pilgrim's Progress* [1676] is that life as a Christian brings problems comparable to those experienced by Bunyan's pilgrim.) The lighthouse in Virginia Woolf's *To the Lighthouse* (1927) has a fairly obvious symbolic force in the novel, but it would be an unwise critic who stated definitively the one thing that it stood for. Perhaps the lighthouse does stand for the unfulfilled dreams of youth, or masculine aggressiveness – but part of its power comes from its multiple suggestiveness and indirect significance.

Symbols are not limited to literature and art: they are central to all known human cultures. When a woman gets married in white she makes use of the symbolic force of that colour for dress within our culture – a symbolic force that has existed for an extremely long time. Any writer who incorporated this convention in a novel would be taking what we can call a public symbol and adapting it for use within his or her work – just as a film-maker who produced a Western with a hero dressed in black would be challenging another (rather battered) convention.

Thus in James Joyce's story 'The Dead' we feel that the repeated references to snow have a symbolic force. This is partly because snow is referred to so repetitively and suggestively that the reader of the story cannot but feel that there is something significant in the function that snow performs in the story. But it is also, of course, because we naturally associate snow with some things rather than others – especially in countries like Britain and Ireland where extensive falls of snow are relatively rare. Put briefly we can suggest that in 'The Dead' snow stands for or suggests *death*: it is cold, it covers the graveyard, it affects the whole country as death comes to us all, and so on. Now the justification for this interpretation is partly that snow is naturally associated with death, because it is cold like a dead body and because people lost in snow die. But it is partly because Joyce in 'The Dead' draws attention to certain of these qualities and associates them with other references in the story (not least with its title) so as to make these associations clear. If we think of the death of Gerald in the snow in D.H. Lawrence's *Women in Love* we will see, I think, that both Joyce and Lawrence are able to incorporate a public symbol into the private or internal world of meaning of their respective fictional works.

But on occasions a writer will create a symbol that has the meaning and significance that it does have only in the context of one particular work. If we think of a green light, for instance, the natural symbolic associations that it has for most of us today are positive: advance, road clear, healthy environment – all of those extensions of meaning that have accrued from the greenness of nature and from our use of green lights in traffic regulation systems. But it is arguable that none of these references or associations are active in our response to the repeated mention of the green light that marks out the quay by Daisy's house for Gatsby in Scott Fitzgerald's *The Great Gatsby*. What Fitzgerald succeeds in doing in this novel is to create a private symbol, something that has meaning only within the world of the novel. Similarly, it is arguable that few if any of the public symbolic associations of lighthouses are active within Virginia Woolf's *To the Lighthouse*; the significance of the lighthouse in that work (and by this I mean the significance for the reader – which is not necessarily the significance it has for the characters in the novel) is something that has to be decided upon on the basis of internal rather than public information.

Another set of terms for public and private symbols is *motivated* and *unmotivated* symbols. In practice, of course, it's hard to find a completely public or a completely private symbol. I suggested, for example, that E.M. Forster's use of the motor-car in *Howards End* developed associations that 'motor-car' might not have in people's minds prior to their reading the novel. But it is clear that for *some* people the motor-car *might* have had these associations, and that the symbolic force that the motor-car has in *Howards End* is, as it were, a potential force already implicit in the motor-car as it was experienced in Edwardian England. (Think of the rôle played by the motor-car in a work of fiction that is near-contemporary to Forster's: Kenneth Grahame's *The Wind in the Willows* [1908]; messrs Toad and Wilcox may belong to very different fictional worlds but they have certain things in common. . . .)

Before concluding I should point out that although I have chosen to discuss things or objects which have a symbolic force, actions and settings can be equally well possessed of this. When Hester in Nathaniel Hawthorne's *The Scarlet Letter* (1850) undoes her hair and lets it fall free, the action has an enormous symbolic charge: in a novel so concerned with repression and concealment it suggests a breaking of bonds, a triumph of natural and healthy impulse over artificial and corrupt restraint. (A friend of mine once described this as the most erotic scene in literature in English!) Similarly, innumerable novels also make use of another symbolic potentiality involved in this scene in *The Scarlet Letter*: the contrast between indoors (representing society, the artificial, restraint), and outdoors – especially in a wild or undomesticated setting (representing pre-social impulses, the natural, the outpouring of feeling). Think of the rôle performed by windows as the dividing-line between these two sets of associations in both Brontë's *Wuthering Heights* and Joyce's 'The Dead'.

In passing from symbols to images I am in a sense going backwards, as an image is arguably a less complex form. I find, however, that 'image' is typically used in a looser sense and is easy to define in relation (and contrast) to 'symbol'.

I referred earlier to Katherine Mansfield's short story 'The Voyage', and commented upon the movement from dark and cold references at the start of the story to warm and – particularly – light images at the end of the story. Let me quote a little from the end of the story:

> On the table a white cat, that had been folded up like a camel, rose, stretched itself, yawned, and then sprang on to the tips of its toes. Fenella buried one cold little hand in the white, warm fur, and smiled timidly while she stroked and listened to Grandma's gentle voice and the rolling tones of Grandpa.
>
> A door creaked. 'Come in, dear.' There, lying to one side of an immense bed, lay Grandpa. Just his head with a white tuft, and his rosy face and long silver beard showed over the quilt.

You will notice the recurrence of words that connote lightness and warmth here ('white', 'warm', 'silver', 'rosy'). These I would dub *images* rather than symbols. The distinction is not an easy one to explain, and there are differences of usage that complicate matters, but the following points are probably worth remembering.

1 Images are usually characterized by their evocation of *concrete qualities* rather than abstract meanings; images normally have a more

sensuous quality than symbols – they call the taste, smell, feel, sound or visual image of the referred-to object sharply to mind.

2 Symbols, in contrast, because they *stand for* something other than themselves bring to mind not their own concrete qualities so much as the idea or abstraction that is associated with them.

Thus in the extract from Katherine Mansfield the feel of the cat's fur, its whiteness and warmth, are brought sharply to our minds *in themselves* – we do not automatically wonder what they 'stand for' (note how this tactile sense is encouraged by the contrast with Fenella's cold hand). But we do not experience sharp sensory responses to Gatsby's green light or Woolf's lighthouse: it is what these stand for or call to mind *apart from* themselves that is important. With Joyce's snow and Forster's motor-car we may pause: I think that in these cases we think both of what they stand for but also of their individual sensory qualities. In these two cases, therefore, we may guess that the references have an imagistic function alongside their primary symbolic purpose.

This is not to say, incidentally, that images do not contribute to thematic elements in a work. Although it is true that an image is distinguished by its concrete qualities in an immediate sense, it sets up waves of association in the mind that have other than a purely concrete significance. Thus, as I have already suggested, the images in Mansfield's 'The Voyage' contribute importantly to our sense of the multi-levelled nature of Fenella's voyage, away from unhappiness and suffering and to the promise of something more cheerful and enjoyable.

Speech and dialogue

According to Bakhtin:

> the decisive and distinctive importance of the novel as a genre [is that] the human being in the novel is first, foremost and always a speaking human being; the novel requires speaking persons bringing with them their own unique ideological discourse, their own language.[23]

This does not, however, mean that people in novels speak the same way as do people in the everyday, extra-fictional world. But it does mean that novels tend to have not one centre of authority – the narrator's or author's voice – but many such centres, centres which typically are in conflict with one another. For Bakhtin, it will be perceived, a voice is not just a

[23] M.M. Bakhtin, 'Discourse in the Novel', p. 332.

mechanical means whereby thoughts are broadcast, it has an ideological dimension. Different voices in the novel represent and disseminate different points of view, different perspectives. And for Bakhtin different voices can be isolated even in a narrator's or a single character's words: when we speak, our utterances contain a range of different voices, each of which carries its own values, such that an utterance can represent a veritable war of different viewpoints and perspectives.

Modern readers are fortunate in that they live in an age of tape-recorders. They can thus study normal, unselfconscious conversation in a way that our ancestors could not – and they can thereby discover that nobody actually talks quite as people are portrayed as talking in novels. The conventions that govern the speech that is represented in novels cover both technical matters such as the syntax of the sentences that make up separate utterances, but also matters such as the sort of things that can be talked about and the sort of language that can be used in such discussions – which have, of course, a significant ideological dimension. There is an amusing exemplification of this point in Josef Skvorecky's novel *The Engineer of Human Souls* (first published in Czech 1977) in which a character, Mrs Santner, tries to defend the 'bad' language used in a novel by referring to how people in real life would actually speak if in the same circumstances as the novel-character in question:

'You have to read it in context,' she explains. 'After all, those words are spoken by an anti-Nazi revolutionary who's afraid the Gestapo will catch him. When people are afraid, they use strong language. It's a well-known phenomenon.'

Her defence is too scholarly for Mr Senka. 'But this is a book, madame,' he cries. 'A book.'

This time it is Mrs Santner's turn not to understand. For her the word 'book' has none of the sanctimonious overtones it has for Mr Senka, for whom a book is a household object to be taken up only on very special occasions. Mrs Santner's husband leans over her shoulder and whispers, 'Now, Betty, don't let's get into an argument,' but this only goads her on.

'I can't help it. In context, language like that has a valid function. That's the way people actually talk in situations like that.'

'But they don't talk like that in *books*.'

Each is partly right, according to his experience.

This is the sort of conflict towards which the realist imperative almost inevitably impels novelists and their readers. We want the novel to give us our recognizable, everyday world, warts and all: but that world includes taboos, prescriptions, repressions from whose attack the novel is not immune. More specifically, we can note that just as writing represents only parts of speech (think how much is lost from a conversation when it is transcribed into writing), so too when speech is presented in fiction further conventions come into operation, such that, although the speech we read in a novel may seem 'realistic', it actually diverges in many ways from actual speech.

One of the extraordinary achievements of realism (and not just in the novel) is that it gives us something that to us resembles the world even though it is formed and constrained by conventions of representation different from those that operate in the real world. Talking about the visual arts, Ernst Gombrich has noted the paradox that the world does not look like a picture, but a picture can look like the world.[24] So far as the novel is concerned, we do not talk like people in books, but the dialogues in books seem to us to be like the conversations we have in real life. As Skvorecky's narrator notes, perceptively, both Mr and Mrs Santner are partly right. Why is this?

The answer has to be that the novelist follows conventions in the representation of speech and dialogue with which we are so familiar that we are unaware of any conventionality. (Just as individuals from Britain or the United States are unaware that they follow conventions governing the nodding and the shaking of heads to mean 'yes' and 'no' – until they travel to a country like Turkey or Bulgaria where these conventions are reversed.) People in novels tend to talk in complete sentences, with few indicated hesitations, mistakes of grammar, 'ums' and 'ers', and so on.

The novelist has to convey exclusively in words what in ordinary conversation we convey by words, tone of voice, hesitations, facial expression, gesture, bodily posture – and by other means. Learning how to do so was not accomplished overnight, and we can note a great difference between the way novelists in most of the eighteenth century represented dialogue and the way later novelists have done so. If, for example, you open Henry Fielding's novel *Joseph Andrews* (1742) at Chapter 5, which is the chapter directly parodying Richardson's *Pamela* in which Lady Booby attempts to seduce her servant Joseph much as Mr B– in Richardson's novel had attempted to seduce Pamela, then you will notice something odd about the layout of the page. Although conversation takes place all through this short chapter, the prose is set out in one continuous unparagraphed stream. Thus Fielding has to keep including

[24] E.H. Gombrich, *Art and Illusion* (New York, Pantheon, 1960). Gombrich explores this paradox throughout the pages of this book.

'tag-phrases' such as 'he said' and 'she replied'. The result is not just that reading the chapter is rather hard work, but that the guiding presence of the narrator keeps intruding: we have narrative tag-phrases in addition to the actual words spoken by the characters.

If we move to Jane Austen's *Pride and Prejudice* we see a very different picture. Dialogue is presented in a recognizably modern form, with each new utterance by a different character given a new paragraph. Here the narrator may intrude or remain hidden at will. If necessary the characters can be left to speak for themselves with no interruption from anyone. This certainly increases the *dramatic* effectiveness of scenes involving dialogue; we feel that we are actually witnessing conversations taking place rather than being instructed by an intrusive stage-manager who keeps pointing out what we have to notice.

The narrator can now use the different possibilities available to create an appropriate effect. Take the conversation between Mr Bennet and his wife that we are given on the first page of *Pride and Prejudice*:

> 'My dear Mr Bennet,' said his lady to him one day, 'have you heard that Netherfield Park is let at last?'
>
> Mr Bennet replied that he had not.
>
> 'But it is,' returned she; 'for Mrs Long has just been here, and she told me all about it.'
>
> Mr Bennet made no answer.
>
> 'Do not you want to know who has taken it?' cried his wife impatiently.
>
> 'You want to tell me, and I have no objection to hearing it.'
>
> This was invitation enough.

Note how Jane Austen wrings so much significance out of her use of Direct and Indirect Speech here. 'Mr Bennet replied that he had not' must be one of the most economically sarcastic lines in English literature: the shift to Indirect Speech somehow conjures up Mr Bennet's weary, long-suffering response to his wife's importuning. We can see that the narrative comments here are more like touches on the tiller than fuller-scale intrusions; we feel that we are witnessing a real conversation but with someone beside us whispering in our ear comments concerning the participants in the discussion.

Ivy Compton-Burnett's *A Family and a Fortune* (1939) shows what is perhaps near to the ultimate rôle that dialogue can play in a novel. At a rough estimate 75 per cent of the novel consists of dialogue, and this dialogue performs a crucial narrative function in the work. The novel's heavy reliance upon dialogue is fascinating and revealing, but the final impression is of a writer rather straitjacketed by her narrative technique. Recent narrative theorists have pointed out that the writer of fiction has at his or her disposal a range of more or less distanced means whereby characters' speech can be rendered. In her *Narrative Fiction* Shlomith Rimmon-Kenan reproduces a useful seven-element division of the full continuum of possibilities open to the writer of narrative fiction. At one extreme we have what she calls *diegetic summary* (in her usage 'diegetic' means loosely 'of or within the story', for which I prefer the more straightforward term 'fictional'.) This is where the reader is given the 'bare report that a speech act has occurred, without any specification of what was said or how it was said'. Second there is *summary, less purely diegetic*, which 'to some degree represents, not merely mentions, a speech event in that it names the topics of conversation'. Third is *indirect content paraphrase (or: indirect discourse)*, 'a paraphrase of the content of the speech event, ignoring the style or form of the supposed "original" utterance'. Fourth is *indirect discourse, mimetic to some degree*, which is a 'form of indirect discourse which creates the illusion of "preserving" or "reproducing" aspects of the style of an utterance'. Fifth comes *free indirect discourse*, which I have already discussed (and which, we should remember, can be used to represent more than just speech). Sixth is *direct discourse*, in which the actual words spoken are 'quoted', although, as Rimmon-Kenan points out, always with some degree of stylization. And finally there is *free direct discourse*, which is 'direct discourse shorn of its conventional orthographic cues', a good example of which would be first-person interior monologue.

If these descriptions are hard to follow, the examples provided by Rimmon-Kenan in her *Narrative Fiction* may make the distinctions more easy to comprehend. All are taken from John Dos Passos's trilogy *U.S.A.* (1938).

1 When Charley got a little gin inside him he started telling war yarns for the first time in his life.

2 He stayed late in the evening telling them about miraculous conversions of unbelievers, extreme unction on the firing line, a vision of the young Christ he'd seen walking among the wounded in a dressingstation during a gas attack.

3 The waiter told him that Carranza's troops had lost Torréon and that Villa and Zapata were closing in on the Federal District.

4 When they came out Charley said by heck he thought he wanted to go up to Canada and enlist and go over and see the Great War.

5 Why the hell shouldn't they know, weren't they better off'n her and out to see the goddam town and he'd better come along.

6 Fred Summers said, 'Fellers, this war's the most gigantic cockeyed graft of the century and me for it and the cross red nurses.'

7 Fainy's head suddenly got very light. Bright boy, that's me, ambition and literary taste. . . . Gee, I must finish *Looking Backward* . . . and jez, I like reading fine, an' I could run a linotype or set up print if anybody'd let me. Fifteen bucks a week . . . pretty soft, ten dollars' raise.[25]

Earlier I used the word 'continuum', and it is worth stressing the fact that although many theorists quantify the possible variations according to grammatical distinctions, the novelist has a wide sweep of alternative possibilities from which to choose, ranging from a summary by the narrator which merely reproduces the gist of what a character has said (and not how it has been said), at one extreme, to a rendering which appears to have no narrator involvement and maximal mirroring of the character's speech both in terms of content and of delivery at the other extreme.

As to *why* a novelist should want to choose to render speech at one point along this continuum rather than at another, a number of different explanations are possible. In one sense we are back to life and pattern; at the number 7 end of the continuum we have maximum life, we are close to the actuality of a character's living use of language. At the number 1 end we have at least maximum potentiality for pattern: the novelist can choose what lessons to draw, what moral to underline, as the character's speech is summarized. Here the novelist's or the narrator's view of the character predominates over the character's actual speech. You can probably think of many other reasons why a novelist may or may not wish to render a given character's use of slang, colloquialism, or dialect. These evoke ways of life, values, cultural specifics which may or may not be what the novelist wants to bring to mind at a given stage of a narrative.

---

[25] Shlomith Rimmon-Kenan, *Contemporary Poetics*, pp. 109–10. Rimmon-Kenan takes her categories and the examples from Brian McHale, 'Free Indirect Discourse: A Survey of Recent Accounts', *Poetics and Theory of Literature* 3, 1978, pp. 249–87.

And, as Bakhtin has pointed out, a voice is not just a 'medium' or a channel of communication, it involves complex ideological elements as well. Take the following passage from Lewis Grassic Gibbon's *Grey Granite* (1934), the final part of the trilogy *A Scots Quair*. Ewan Tavendale is talking to a number of other apprentices, and one of them, Norman, reveals that, while he can never be bothered to go to a trades-union meeting, his father has a more positive attitude:

And he looked a bit shamed, *He's Labour, you see.*

Tavendale said *There are lots of chaps that, my stepfather was*, and you all cheered up, sitting on buckets in the furnace room, a slack hour, and having a bit of a jaw, you were none of you Labour and knew nothing about politics, but all of you had thought that the Bulgars of toffs were aye Tory or Liberal or this National dirt. And somehow when a chap knew another had a father who'd been Labour you could speak to him plainer, like, say what you thought, not that you thought much, you wanted a job when apprenticeship was over and a decent bit time and maybe now and then a spare bob or so to take your quean to the Talkies – och, you spoke a lot of stite like the others did, about the queans that you'd like to lie with, and the booze you'd drink, what a devil you were, but if you got half a chance what you wanted was marriage and a house and a wife and a lum of your own. . . .

Now it hardly needs saying that on a simple level the use of elements from Scots dialect gives this passage added verisimilitude. If we are not Scottish then we may be sent to a specialist dictionary to discover the meaning of some of the dialect terms in the passage, but most of us can probably guess at their meaning without too much difficulty, and British readers will probably work out that 'Bulgar' is a euphemism for 'Bugger' – a term that Gibbon's publisher would certainly have bridled at in 1934.

But, beyond vocabulary, the language of the passage is crucial to certain other effects. We can say that it mimics the speech and thought patterns of a working man of the period, especially one talking among a group of his peers. The use of 'you' is especially interesting. On the one hand it represents an accurate piece of observation on Gibbon's part; in Britain men of this background will use 'you' as a generalizer where a person from a middle-class background would use 'one' (as Virginia Woolf does throughout her writing, and the British Royal Family do throughout their speech). But more than this, it involves the reader in a process of collective sharing of attitudes: we feel as we read that we are part of the group that is thinking out its beliefs, it is as if these beliefs come out of a group to which we belong. We thus experience the ideology of the group from the inside while having it displayed in a manner that allows us to look at it more objectively. The natural form of the work-group conversation conceals what is actually quite an unnatural process – whereby a group is made to speak those taken-for-granted items of ideological positioning that, just because they *are* taken for granted, are normally not spoken. (Virginia Woolf does something rather similar in *The Waves*, although revealingly the statements in this work are more individual and individualistic; the class with which she is concerned lacks the collective voice and consciousness that Gibbon wishes to display.)

There is, in other words, a sort of sleight of hand in the above passage. Gibbon is actually in one sense telling us something about the working class from a vantage point outside of its typical self-knowledge. But, because he dramatizes this telling in the form of a discourse that belongs to the class, the reader is given the impression that as we *experience* the ideology with the collectivity, so too we are simultaneously detached from it and allowed to look at it through its members' self-aware analysis of themselves. (One of the things that is being mimicked here is the style of a working-class anecdote. It is as if the reader is being addressed in an intimate way by the collective voice that is thus personified through the language that it speaks.) Gibbon could not have achieved such a complicated effect without having borrowed the language and habits of expression of the group, even if he uses these in a way that members of the group would not have done.

We should, in other words, be very suspicious of views which look upon the language that a novelist uses to represent speech and dialogue (or thought, or unexpressed identity even), as merely a medium. A medium is relatively untouched by what it transmits. Bakhtin has argued that 'the language of a novel is the system of its "languages"'[26] – in other words, that the action that takes place on the surface of a novel is only one level of the total action to be found within its pages. Below the level of literal event is the level of ideological action, an action in which the participants are not characters but positions encapsulated in language. For Bakhtin:

> The word in language is half someone else's. It becomes 'one's own' only when the speaker populates it with his own intention, his own accent, when he appropriates the word, adapting it to his own semantic and expressive intention. Prior to this moment of appropriation, the word

[26] M.M. Bakhtin, 'Discourse in the Novel', p. 262.

does not exist in a neutral and impersonal language (it is not, after all, out of a dictionary that the speaker gets his words!), but rather it exists in other people's mouths, in other people's contexts, serving other people's intentions: it is from there that one must take the word, and make it one's own.[27]

This, surely, is what we see in the passage from Gibbon: men using other people's words, words which contain opinions that are not their own but which they are led to believe *are* their own, while they begin to discover their own views and interests, and begin to appropriate the word for their own uses.

[27] M.M. Bakhtin, 'Discourse in the Novel', p. 293.

7 Studying the Novel

To study the novel means, in part, being conscious in your reading of some of the things that I have been talking about in this book up to now, but it does not just mean this. We may read a shortish novel in one sitting, while anything up to twenty or more sessions will be required for a long and complex novel such as Tolstoy's *War and Peace*. For those readers who read Victorian novels when they were first published in serial form there was no choice: they *had* to stop and wait at certain stages in their reading. When we pause in our reading of a novel we go over what we have read and we think forward to what we guess will happen or what we would like to happen. Frequently we will imagine ourselves in situations described in the novel; perhaps we will hold imaginary conversations with characters, or wonder what we might have done in their positions.

Put another way: expectation, surprise, disappointment, foreboding, tension, suspense, imagination, fantasy – all form part of our reading of a novel. To read a novel is to be involved in a *cumulative process*.

The problem which this raises for the study of the novel is that there are certain experiences in us, and certain events in the novel, which lodge far more permanently in the mind than do others. Indeed, it is part of the way that a novel works upon us – part of its power as a source of enjoyment – that some aspects of our reading experience should fade more quickly than others. Moreover, our memory plays tricks with us once we have finished a novel. If it is some time since you read *Wuthering Heights* then try this test on yourself: at what stage in the novel does the older Catherine die? I suspect that unless you are an unusually retentive and conscientious reader you are likely to place this event far later in the novel than it actually occurs.

The issue is complicated by the fact that we often read a novel more than once. One of the distinguishing features of good novels (unlike pulp fiction) is that we can get something new out of a second or third reading of the same piece of fiction. On a subsequent reading we are normally less preoccupied with what will happen and so are able to read more carefully and notice many details that slipped our attention on first reading. So to *study* a novel we need to find ways of preserving things that would otherwise be lost.

In ordinary life when we want to preserve something we use aids: holiday photographs, diaries, memo-pads, and so on. And to study the novel we have to do the same. We need to take notes. Taking notes is actually a skilled operation and one which one has to practise in order to be good at it. And it takes time before one can develop one's skill to the point at which it does not interfere with one's reading. So here are some suggestions.

How to take notes

One can write notes either in the novel one is reading or in a separate notebook. The advantage of the former method is that it does not disturb one's reading too much; the disadvantage is that it spoils a book, it affects one's second reading of the novel, there is not always sufficient room for notes in the book, and *retrieval* of the notes afterwards for purposes of study is difficult. One way round these problems is to write very brief notes in pencil in a novel as one is reading, and then to copy these up and expand them in a separate notebook or folder later on. This also allows you to copy out brief extracts from the novel which strike you as important, and it means that you file for future reference only notes about which you have thought a second time after having finished reading the novel. Pencil marks can be erased once they have served their purpose.

If something in the text strikes you as important or significant but you don't want to puzzle over why lest you break your train of thought, then mark or underline the passage in question, note the page reference on the inside back cover of the book, and come back to ponder the point once you have finished the novel.

What to note

Learning what is of significance in a novel, and in your response to it, is a matter of practice. Teachers and lecturers often ask for analyses of selected passages from novels, and undertaking such analyses is an excellent way of sharpening your eye for important detail in the reading of fiction. Such close reading is, however, only part of what constitutes a full critical reading of a novel, for in addition to responding to significant detail in the prose of a work of fiction you need also to be able to perceive larger patterns and movements in the work as a whole, as well as connections to other works. The list that follows is intended to serve a double purpose. On the one hand I hope that it will be useful to work from when you are engaged in the analysis of a particular passage from a novel or a short story. In addition, however, I hope that its perusal will serve as a reminder of the sort of points that can be noted in the process of reading or rereading a work of fiction. The golden rule here is: if in doubt, then make a note.

Checklist

- *Narrative technique*: All information relating to the manipulation of narrative in the work: clues about the values or personality of the narrator; voice and perspective; what the narrator *doesn't* know; changes of narrator or narrative perspective; narrative intrusion or comment. 'Telling' and 'showing'. Differences between story and plot.
- *Tone*: Is it familiar or formal, intimate or impersonal? Who (if anyone) is apparently being addressed? Do the vocabulary or syntax suggest a particular style of delivery?
- *Characterization*: Information about *how* we learn about characters; any indication that characters are changing or developing; significant new information about a character; views as to what the writer is trying to achieve in the presentation of character.
- *Speech and dialogue*: Use of Direct, Indirect or Free Indirect Discourse. Do characters speak for themselves or does the narrator intrude, comment or direct? Is the dialogue realistic or conventional? What functions does it perform? (Development of character, of plot, introduction of dramatic element, discussion of theme[s].)
- *Thoughts/mental processes*: Do we 'get inside characters' heads'? If so, which heads, and how?
- *Dramatic involvement*: Is the reader drawn into events as they happen, or rather encouraged to observe them dispassionately? How is this achieved? (Manipulation of distance.)
- *Action*: Any information that advances the plot, gives significant new developments in human relationships or new events.
- *Setting and description*: What is significant about where the action takes place, and about descriptions of people and places?
- *Symbol or image*: Anything apparently significant should be noted. Do symbols/images relate to others used elsewhere in the work?
- *Theme(s)*: Any development of themes dealt with elsewhere in the work; introduction of new thematic elements. Moral problems/issues raised for the characters or for the reader.
- *Your own response*: Strong personal preferences/responses the work evokes in you – or dislikes/disapproval. Mood. Strong identification with a character – or the opposite. Tension, desire to know what happens. Particular expectations (especially at the end of chapters/sections, or when a mystery or problem is presented). Any experience of bafflement

or surprise; and points where you feel you disagree with or react against a narrative opinion (or the opinion of a character).

Revision

If you have taken full notes from your reading of novels and your study of critics then the process of revision should be easier for you. Always study your notes in conjunction with the novel concerned; it makes no sense to memorize a set of notes if you have forgotten the salient details from the work itself.

You may take an examination requiring knowledge of a number of novels up to a year after you have read some of them. Full rereading is impossible, and some elements will certainly have slipped your memory. *Don't* try to skip-read the whole work again (unless it is a short story), but concentrate upon reading selected passages with care. These should always include the opening and closing pages of the work – beginnings and endings of novels are invariably revealing. Is the opening dramatic or descriptive? Does it plunge the reader into the middle of things, or carefully establish a scene? Does it set a dominant tone, and establish the narrative perspective or personality of the narrator? Is the reader addressed directly, and are we given an indication of what sort of reader (or reading) the narrative invites or appears to assume? Is the ending happy or sad (and why does the author choose to make it so?)? Does it tie up all the loose ends or leave many questions unanswered? (And with what effect?)

Apart from opening and closing pages, you should pick some key passages – either those that have seemed important to you, or those which critics have found of great interest. And pick a couple of passages at random, preferably ones that seem unfamiliar on leafing through the work. Analyse these in detail.

If it makes you feel easier then construct a list of character names (just the main characters) for each work prior to an examination. But if you forget a name in an examination don't panic, just leave a blank and go on writing. If the name doesn't come to you by the end of the examination then asterisk the gap and write in a footnote explaining that you have forgotten the character name and indicating who the character is by other means such as relationships to other characters.

Essays and examinations

When it comes to writing essays and examination answers there is no substitute for practice: the more that you write and have your writing commented upon, the more clearly and easily you will come to write. I am suspicious of student guides that treat essay writing rather like learning how to use a computer program: some examples almost give a standard essay with blanks for the student to fill in with the appropriate words taken from the topic. The following remarks are, therefore, tentative and meant to help you think about how you prefer to write. I have attempted to give some general advice and also to focus upon some of the specific problems involved in responding to questions about the novel. But different teachers, different colleges, and different cultures have different conventions governing the writing of essays and examination answers. Listen to the advice of your teachers, take note of their comments on your work, and ask them if you are in any doubt. Above all, practise. It is generally better to write ten two-page essays in a term than one twenty-page essay (and much less boring). That way you learn to think about how to respond to questions and how to structure answers.

With this general disclaimer in mind, read the following suggestions and consider the extent to which they represent good advice. Are there things you disagree with? Does anything below contradict other advice you have had? What have I left out?

- *Be relevant*: You will be responding to a specific question or topic. Do not just pick out a couple of words from it and proceed to write all you know about the text(s) mentioned. (This is known as the sack-of-potatoes method: you spend months gathering in the potatoes, and then the first time you eat you slit the sack open and spill them all out all over the table. Remember that an essay or answer is like a meal: selection, preparation and order are necessary if the result is to be pleasing.) Use coloured pens and underline key words in the question. If the question has several points then *number* these. After every second page you write, look back to make sure that you are being relevant.
- *Argue a case*: A question requires an answer, not just a response. It is expected that you take up a position to what has been asked, and argue your position as logically as possible. Do not be afraid to express a personal opinion, but, if you do, say that it is a personal opinion and justify it (see below). If you refer to the opinion of others (critics, for example), say whether or not you agree with them, and why. If the question or topic includes a quotation, remember that you may disagree with the opinion of the person quoted. (Very often the person who has

written the question expects you to do this.) But, if you are to argue successfully, you need to . . .

- *Plan your answer*: If you only have a very short amount of time, you may have to restrict this to a five-minute sketch of what you intend to say, but if you are writing an essay in your own time then you should plan more carefully. With an essay, you can afford to restructure what you have written for its final presentation, such that the introduction and conclusion can be added on at the end in case your argument has become modified or more complex in the course of writing. (This is much easier if you use a computer: even if it means pawning your hi-fi, you should try to get access to word-processing facilities. They will allow you to write better essays in half the time. In particular, you can shift paragraphs around and re-order them before your final rewriting so that the argument unfolds logically.)

  If, as is often the case in Britain and the United States, you are faced with writing three or four examination answers in three hours – a situation which my students in Norway (who typically have six or eight hours to answer one or two examination questions) find hard to accept is possible, you may find as you write that your examination answer is moving in a rather different way from that laid out in your introductory paragraph. If so, the best thing is to admit to this change of direction, and even to explain why.

  A good essay or examination answer should have an introduction which explains what attitude you are taking to the question asked, and how you are going to answer it. This need only be half a page long. You should then provide, in as ordered a way as possible, evidence to substantiate the position you have taken up. At any stage in an essay or examination answer the person reading it should know (i) what you are trying to argue, and (ii) how this argument fits in to your general thesis or response to the question. Anyone who has marked examinations knows that a sure sign that there is something wrong is when you have to flip back to page 1 to check exactly which question the script is supposed to be answering.

- *Back up your arguments with evidence*: There are different sorts of evidence that can be adduced in an essay or an examination answer. Unless the question specifically focuses upon extra-textual matters, the most important evidence that you can bring forward is likely to be *textual*. The general rule about textual evidence is that it should be *detailed* and it should be *analytical*. In an essay you should quote enough to make your point, and you should *never* assume that a quotation speaks for itself. Every time that you quote from a novel or a short story in an essay you should explain what it is in the quotation that is important –

pointing to *particular* words, *particular* phrases, *particular* techniques (such as those outlined in my suggested checklist).

This is especially important in writing on the novel, because there is a tendency to assume that the language of a novel speaks for itself. The language of many novels seems almost transparent – we look through it at 'what happens' and are not arrested by it *as language* as we typically are when reading a poem. You have to learn to resist this process and to 'look at' that which in the past you have 'looked through'.

Examinations may present a different problem. If your college or university is an enlightened one you may be allowed to take unmarked copies of texts into examinations. Generally speaking, examinations have to be written without access to the novel or short story texts on which they are based. I do not recommend that you spend hours learning selected passages by heart. Not only is this extremely time-consuming, there is also a tendency for those who have shed blood learning long quotations to insist on demonstrating the fact in the examination room, and the quotations that have been learned may well not be relevant to the question asked. But, even if you cannot quote at length from a novel, you can still be detailed in your reference to it. Refer to particular *scenes*, isolate the operative *events* and *actions*, and *utterances* made (you do not have to quote to do this), draw attention to relevant aspects of *narrative technique* observable, and comment upon such things as *symbolism and imagery* if they play a significant rôle. If you can refer to particular words and phrases used, well and good.

Apart from textual evidence, there are various forms of extra-textual evidence that can be incorporated in an answer: comments and opinions from critics, statements from or about the work's author, parallels with other works either by the same author or by others, information of the sort mentioned in my brief discussion of 'the sociology of literature' (p. 166), information about relevant social, historical, cultural events or states of affairs. In all of these, two issues should be borne in mind: *influence* and *relevance*. You may reproduce a lot of accurate facts, but did they have any influence on the writing (or do they have any influence on the reading) of the work? Are they relevant to the question asked?

- *Do not tell the story*: I suppose that I ought to concede here that it is possible that one day you might encounter an essay topic or examination question which starts, 'Tell the story of *Wuthering Heights* and then proceed to comment on . . .'. This apart, 99.9% of essay topics and essay questions do not require you to open with an obligatory three-page précis of what happens in the novel in question. You have to assume that the person marking the essay or answer has read the work and knows what happens: he or she does not need or want to be reminded of this.

    This is a very common flaw in responses to questions about prose fiction (and drama). You are a literary critic not a journalist: your job is to comment on the novel, not to reproduce it in abridged form.

● *Remember that the characters are not 'real people'*: Although it is quite legitimate at times to discuss the characters and actions in a novel as we discuss the people we know in real life (all critics do this at some time or another), you should also indicate clearly that you are aware that what we read are words that have been created and crafted by an author. Peter Lamarque has suggested that the reader of fiction needs to be able to combine the imaginative involvement of an internal perspective with the awareness of artifice of an external perspective,[1] and you should make sure that the person reading you knows you are aware of this difference.

Perhaps I can conclude by attempting to exemplify some of the things that I have been recommending. Imagine that you are writing a response to the following question: 'Discuss Kingsley Amis's humorous techniques in *Lucky Jim*, drawing attention to the stated or implied values that underpin them'. Imagine too that you have decided to use the following quotation from the second page of the novel as a basis for your comments. The anti-hero of the novel, the university assistant lecturer Jim Dixon, is talking to his professor, Welch. Welch starts off describing a musical evening that had taken place at his house, 'with a tremolo imparted by unshared laughter':

'There was the most marvellous mix-up in the piece they did just before the interval. The young fellow playing the viola had the misfortune to turn over two pages at once, and the resulting confusion . . . my word . . .'

    Quickly deciding on his own word, Dixon said it to himself and then tried to flail his features into some sort of response to humour. Mentally, however, he was making a different face and promising himself he'd make it actually when alone. He'd draw his lower lip in under his top teeth and by degrees retract his chin as far as possible, all this while dilating his eyes and nostrils. By these means he would, he was confident, cause a deep dangerous flush to suffuse his face.

    Welch was talking yet again about his concert. How had he become Professor of History, even at a place like this? By published work? No. By extra good teaching? No in italics. Then how? As usual, Dixon shelved this question, telling himself that what mattered was that this

---

[1] Peter Lamarque, *Fictional Points of View*, p. 14.

man had decisive power over his future, at any rate until the next four or five weeks were up. (Ellipses in original)

Here is my stab at a brief answer that does everything wrong in terms of the advice given above:

Lucky Jim by Kingsley Amis is the story of a young history lecturer who is scared of losing his job. He tries to make his professor like him and want to keep him in his department, but he does everything wrong. He gets drunk and burns the sheets at his professor's house, falls in love with the professor's son's girlfriend, and gets drunk again giving a public lecture. But everything comes out all right in the end.

Amis makes him think lots of funny things which make us laugh. Also the tight spots he gets into are funny.

This passage shows us how laughable the professor is. Jim also imagines funny things here, like pulling a silly face. I think that anyone reading this passage will see how funny it is. This shows that Amis likes Jim and doesn't like Professor Welch. Perhaps Professor Welch did something nasty to Mr Amis once.

I like this book, and the passage shows why.

And here is the sort of answer I personally prefer to receive:

Humour is often considered a personal matter, yet most readers find *Lucky Jim* a very funny novel. Such agreement – and agreement about the humour of particular passages, suggest that it should be possible to analyse why and how the novel is funny. My analysis will take a short passage from early on in the novel and will attempt to isolate techniques in it that operate throughout the work. At the same time, I will attempt to suggest what values lie behind the techniques by analysing the way in which they work on the reader.

What strikes us straight away if we look at this passage is that, whereas Professor Welch's speech is quoted but his private thoughts are not revealed (if he has any here), Jim Dixon says nothing out loud but the reader is given access to a succession of his unexpressed thoughts. One obvious source of humour is the use of sharp *contrast*, between what Jim is thinking and his public appearance. The contrast is backed up by verbal wit – 'Quickly deciding on his own word' allows the reader to feel in alliance with Jim; the suggestion is that because we think along the same lines as he does we do not need to be given Jim's 'own word'.

This is a pattern that continues throughout the novel up to the dramatic point, in Jim's final confrontation with Bertram, when Jim thinks something to himself and then, for the first time in the novel, utters the same words publicly. We note that Welch's tremolo is 'imparted by unshared laughter'; he is wrapped up in himself, unaware of Dixon's lack of interest in his tedious stories. He is not just absurd, he is also objectionable: behind the humour lies a revulsion against his self-importance and lack of interest in Jim and his views. Amis's use of ellipses at the conclusion of Welch's speech is indicative of the character's mindless self-obsession: the recalled scene is amusing to him and he does not feel impelled to explain why it is amusing to his companion; because he is really only reliving the experience for himself he judders to a halt without making a point – his egocentricity does not require that verbal precision that would communicate something specific to Jim.

The three quoted paragraphs show an interesting progression from Direct Speech (paragraph 1), through Reported Speech (paragraph 2; the 'speech' is, of course, Jim's inner speech), moving eventually to Free Indirect Discourse (specifically, represented [verbal] thought) by the second sentence of the third paragraph. This is not accidental: in the first paragraph we are put into the position of listener to Welch's banalities, we are outside, being ignored by him as Jim is ignored. In the second paragraph we go inside Jim's mind, but still as an observer, able to contrast the absurdity of his wild and physical imaginings with Welch's tedious and petty ramblings, but by the end of the third paragraph we *are* Jim: the narrated thought recreates Jim's thought-processes in our own reading minds. Thus the reader is encouraged to look at Welch from the outside but to experience with Jim, from the inside. We understand that Jim has to talk to himself, to engage in a question-and-answer dialogue with himself, because proper communication is impossible with a man so self-important and so uninterested in others as is Welch.

I have used the word 'wild'. We can note that there is a wild, physical, anarchic element to Jim's fantasies: Jim is recurrently associated with a free physicality that contrasts with the polite sub-academic drawing-room conventionality of Welch. Note the use of a word such as 'flail' to describe his face-pulling, and the detailed physical description of the actions associated with the planned face-pulling. Physicality in novels often stands for 'natural' as against 'artificial' or 'conventional' behaviour. But Jim's physicality is inner, secret, repressed: thus we may suspect that behind the reader's laughter lies a sympathy with the natural man forced to repress a native honesty and lack of pretension in a context demanding adherence to artificial forms of polite deception. This, we may note, is a standard component in the literary works associated with the so-called 'angry young men' of the 1950s. In the work of all of these, including *Lucky Jim*, there is an anger at certain failings such as hypocrisy, privilege, convention, that may be portrayed as individual failings but which are also seen by their accusers as faults endemic in

British society of the immediate post-Second World War period. We should remember, then, that behind our pleasure at the attack on Welch may be a belief that he represented (and perhaps still represents) a set of social ills from which our age has yet fully to free itself.

In conclusion, we may note that although comedy has often been considered to represent a lower, simpler and more basic form than tragedy, we should not make the mistake of assuming that humour is a matter only of the surface, that what makes us laugh has no depth. We would not laugh as we read this passage if we rejected the values that underpin Amis's descriptions. If we felt Jim to be an outrageous, rude, ungrateful and hypocritical lout who should show more respect to such a decent and lovable man as Professor Welch, then this passage would not make us laugh. (I have met people who do believe this and who do not find the novel at all funny.) It is only by succeeding in making common cause with the reader through a shared set of human values that Amis succeeds in making him or her laugh.

However inadequate you find my preferred version, I hope that at least you will agree that it is superior to the earlier answer.

8 Critical Approaches to Fiction

The following survey is designed to help students in a number of different ways. First, I hope that it will make *recognition* of different critical approaches easier. (By a critical approach I mean both the underlying theoretical assumptions of the critic concerned, but also his or her particular methods of analysis and inquiry.) Such recognition may not always be easy, as, although it is relatively straightforward to describe a critical approach in rather abstract and general terms, in practice particular critics often combine a number of different assumptions and methods in their critical work.

Second, I hope that the following brief survey will help you to think about the variety of ways in which the novel can be read and discussed.

Third, you will notice that not all the approaches outlined below can be simply combined, as there are in some cases contradictions between them. You have to make up your own mind, in such instances, as to what your position is.

Textual approaches

By 'textual approaches' I mean those critical discussions of novels which restrict themselves to information gained from the actual texts of novels discussed. Of course this restriction can never be absolute – we have to take some knowledge of the world to a discussion of any novel in order to make sense of it – but textual critics concentrate on the actual words of the novel(s) they are studying rather than bringing what is called *extrinsic* information into their criticism. Textual critics thus pay little or no attention to biographical information about the author (including other writings by him or her), information about the author's society and historical period, the history of readers' responses to the novel, and so on. The relation between such a textual critic and the text of a novel is similar to that between a fundamentalist Christian and the Bible: whereas some

Christians suggest that the Bible has to be studied in its social and historical context, more fundamentalist Christians tend to read it 'in itself'.

Critics known as *formalists* take a particularly exclusivist attitude to the text. A formalist critic is one who pays great attention to the *form* of a literary work, but the term has been applied mostly to two groups of critics: the Russian and Czech formalists whose most influential work was written in the second two decades of the twentieth century, and the Anglo-American New Critics who flourished in the 1940s and 1950s. The New Critics – to oversimplify somewhat – at least in theory rejected what they termed 'external' or 'extrinsic' information more or less *in toto*, preferring to concentrate upon 'the work itself' in relative isolation. In terms of *method* they often use what has been called *close reading* – taking a small section of a novel and reading this in exhaustive detail, drawing attention to the sort of things that I outlined in my earlier 'checklist' in the previous section. And as the New Critics (unlike the Russian and Czech formalists) developed their critical ideas and practices mainly in connection with the interpretation and analysis of poetry, they tended to bring to the criticism of the novel a new concern with such matters as linguistic detail, paradox, tension, irony, symbolism, and so on. Thus the title of an essay written by the New Critic Robert Heilman and first published in 1948, '"The Turn of the Screw" as Poem', can be seen to announce a representative critical trajectory.

Such critics have, by common consent, contributed enormously to our understanding of the complexity and subtlety of great works of fiction, moving us away from merely discursive discussion of plots and characters and demonstrating that the novel is as receptive to detailed analysis as is poetry. My chapter on 'Analysing Fiction' could not have been written without the work of such critics. Many more recent critics who are by no means formalists have been significantly influenced by the practice of the New Critics. To take one highly recommended example: David Lodge's *Language of Fiction* (1966) is one of the major critical studies of the novel of the past three decades. Lodge's analysis of Charlotte Brontë's *Jane Eyre*, for example, is a masterly isolation of patterns of reference and connotation in Brontë's novel, treating it almost as a poem that can be analysed in terms of its imagery as much as in terms of its plot and characters.

One important thing that such critics have reminded us of is that novels are *made*; they are not windows giving directly on to reality. They have thus led us to be much more wary about discussions of characters and events in novels which treat these as real people and happenings. A key influence here was L.C. Knights's essay 'How Many Children had

Lady Macbeth?' (1933), which was concerned not with the novel but with drama. Knights was by no means a narrow formalist himself, but his essay has made critics aware of the dangers of referring to the life of characters 'outside of the text'.

Generic approaches

A genre is a type or kind of literature, and we often refer to the novel itself as one of the three main literary genres along with poetry and drama. But there are many different ways to classify literature generically, and the novel has been divided into a number of subgenres by various critics – which is what I myself did earlier on in this book when I distinguished between 'Types of Novel'. A generic approach to prose fiction insists that we cannot begin to read or understand a novel until we are clear as to what *sort* of novel it is – which for most critics means getting clear about the author's intentions with regard to his or her work. Thus – the argument goes – unless we understand the picaresque tradition within which Defoe's *Moll Flanders* is written we will run the danger of misreading and misunderstanding this novel.

   This sort of initial orientation, of getting clear about what sort of work one is about to read, can often be one of the most useful benefits of reading critics. It is always a useful question to ask oneself once one is several pages into a novel whether one is clear as to what sort of novel it is. Sometimes an author may want to baffle and surprise a reader, but sometimes a modern reader may need the help of historical scholarship and criticism in order to get clarity on such matters.

Contextual approaches

If the most valuable aspect of the work of formalist critics has been their development of methods of critical analysis, the aspect of their work which has gone least unchallenged by others has been their rejection or playing down of 'extrinsic' information as an adjunct to the reading and criticism of literature.

   Sociological and *Marxist* critics have placed great stress upon the need to understand the context of the author's own society and his or her position within it, both as an individual (a member of a particular social group or class) and as an author (a member of a literary group, relying on certain publishers, libraries, readers, and so on).

   Take Joseph Conrad's *Heart of Darkness*. Recent critics have suggested that the novella can only be fully appreciated in the context of what European powers were doing in Africa in the last part of the nineteenth century, and in particular in the context of statements made by prominent Belgians – including King Leopold – concerning the Belgian exploitation of the Congo (in which Conrad himself had worked). Thus some of Leopold's statements on the (extremely brutal) Belgian exploitation of the Congo include comments about opening to civilization the only part of our globe where Christianity has not penetrated, and piercing 'the darkness which envelops the entire population'. Conrad's ironic use of words such as 'civilization' and 'darkness' in the novella, it has been argued, can only fully be appreciated in the light of his aim to mock and expose statements such as Leopold's.

But some contextual critics argue that more than this is needed. The narrative indirection in the novel – Marlow's inability to tell the truth to Kurtz's 'Intended' and his feeling that he cannot make himself understood to his auditors – mirrors a situation in which comfortable Europeans who had never witnessed colonial brutality at first hand in Africa were unable to hear the truth spoken by such as Marlow. This argument raises some problems in its turn. Conrad deliberately left out localizing information such as the name of the European country or the part of Africa in which the novella was set, and with regard to his short story 'Youth' he made it clear that he very much objected to having such information re-inserted.[1] It has been argued that the novella itself contains as much of this sort of contextualizing information as is needed. Key issues here are: what information points to material that had a *determining* effect on the writing of the work? What information is *relevant*?

This use of extrinsic information may go beyond what an author intended, and it may also go beyond the sort of detailed issue that I have mentioned above. Thus one might argue that much of Henry James's work presents the reader with something akin to voyeurism; seeing, observing, analysing reality is more highly valued than actually intervening in it so as to change things. Such intervention is typically seen as crude, common, clumsy. Now as use of the word 'voyeurism' suggests, one way to explain such a tendency might be by reference to James's own proclivities, especially his sexual ones. But it is equally possible to draw a parallel here with James's social position – the place he occupied in the class structure of both the United States and Britain. At this time the real power of members of the intellectual élite to which serious writers

---

[1] In a letter to Richard Curle: see Richard Curle (ed.), *Conrad to a Friend: 150 Selected Letters from Joseph Conrad to Richard Curle* (London, Sampson Low, Marston & Company, 1928), p. 297.

belonged was waning; they retained a comfortable position in society, but one which was far less influential than it had been. The familiar voyeuristic stance with which we are presented by so many of James's works may, then, reflect not (or not just) James's own psyche, but the social rôle he occupied as a writer. By valorizing observation rather than action James is actually arguing indirectly for the importance of his own social rôle.[2]

Sociological critics have also drawn attention to the important influence that factors referred to collectively as the 'sociology of literature' can have on the creation of works of literature: publishers, patrons, literacy rates, readership profiles, libraries, booksellers, and so on. It is clear that if only a part of society is literate and can read novels, then this will affect the sort of novels that are written – as was the case in eighteenth-century England. And, if success as a novelist depends upon having your work accepted by the circulating libraries, and these do not accept anything that does not conform to a set of rigidly moralistic views – as was the case for much of the nineteenth century in Britain – then the reader should know this before he or she expresses dissatisfaction with the portrayal of sexual relationships in a nineteenth-century novel.

A recent development of some interest has been the emergence of very loose groupings known as cultural materialism and New Historicism. The critics in question take some of the insights (or claims) of post-structuralism (see below) concerning the difficulty or impossibility of confronting or interrogating reality other than through texts of some sort, while retaining some sort of what we can call a realist commitment – that is, a commitment to a belief that reality is not just texts or closed systems of meaning. New Historicists typically juxtapose literary and non-literary texts in ways that produce unexpected and frequently highly illuminating flashes of insight. Cultural materialists are especially interested in following the life of literary works through successive ages and varied cultures, and often responses to and interpretations of a literary work are as much of interest to them as is 'the work itself'. It has to be said, however, that up to now these critics have paid proportionately less attention to the novel than to other literary genres.

---

[2] I expand these arguments in my *Cunning Passages: New Historicism, Cultural Materialism and Marxism in the Contemporary Literary Debate* (London, Arnold, 1996).

Biographical approaches

The more books by the same novelist that we read, the more we become conscious of common or similar elements in them (as well, of course, of contrasts, new directions, surprise variations), and the more we begin to build up a picture of the man or woman behind the printed text, constructing an idea of his or her values, interests, and attitudes.

Such an interest in an author can lead us to want to know more about the person in question in his or her own right, and then we may start to read the novels in the light of what we have learned about the author – even in terms of what he or she has said about them explicitly. Such a biographical approach to the study of the novel – reading fictional works with explicit reference to the life, personality, and opinions of the author – seems straightforward enough, yet it is fraught with problems and has involved many sharp debates. D.H. Lawrence advised: 'Never trust the artist. Trust the tale',[3] and generations of critics have pointed out that novelists *can* be very bad guides to their own work.

Why should this be so? We can isolate a number of possible explanations. First is that of *inspiration*. The writer seems often not to use the conscious and rational part of his or her mind for literary composition – something that has been appreciated since the days of antiquity when the term 'inspiration' was first used. So that secrets of the created work may not be accessible to the rational, enquiring mind in the same way that details of other actions and utterances are. Second is that the author may want actually to *conceal* something: a real-life model, a confessional element in the work, or whatever. Third, the author may comment on his or her work a very long time after writing it.

When it is a matter of relating a novelist's work to his or her life then similar problems may emerge. How do we know which experiences in a writer's life were reflected (or transformed) in his or her work? Experiences which seem minor to us may have been crucial to the person who had them – but, even so, they may not have influenced the writing of a given work.

If all this sounds like a counsel of despair I should add that it is not meant to be: I *do* think that biographical information can be of very great help in enabling us to respond more fully to a novel or a short story. But we should be fully aware of the problems involved in such critical

---

[3] D.H. Lawrence, 'The Spirit of Place', in Antony Beal (ed.), *Selected Literary Criticism: D.H. Lawrence* (London, Mercury Books, 1961), p. 297.

procedures, and we need always to be sensitive to what the work itself can tell us.

Take D.H. Lawrence's novel *Women in Love*. It seems very likely that Lawrence's experiences in the period of the First World War had a crucial effect on the writing of this novel even though the war is not mentioned at all in it (in spite of the fact that it was written in the middle of these experiences). Lawrence was medically unfit for service in the war, and came under suspicion from the authorities for his anti-war views and because his wife Frieda (neé von Richthofen) was the cousin of 'the Red Baron' – a man notorious for shooting down scores of British aeroplanes. Lawrence and Frieda were even forced to leave their cottage on the Cornish coast – allegedly because they were suspected of signalling to German U-boats! Now although none of this is directly referred to in *Women in Love* it is arguable that much of the tone of that novel – in particular its dismissive remarks about England – can only be fully understood in the light of these experiences of Lawrence's. Moreover, the comments in the novel by Birkin about the destructiveness of the 'mechanical principle' have, too, to be seen in the context of Lawrence's complex responses to the slaughter of the war – a slaughter that is not mentioned directly in the novel at all.

I stress that this sort of interpretation has to be handled with extreme tact and sensitivity, and it is a good idea to try to read and respond to a novel independently of such information first. If the information *then* seems to enable one to understand the work better or to clear up puzzles in one's own response then there is a better chance that one's interpretation is not doing too much violence to the work.

Psychological and psychoanalytic approaches

Biographical criticism can often shade off into psychological criticism: trying to use the work to uncover the psychology of its creator, and then to use any insights so gained to shed fresh illumination on the work. Psychological criticism can do more than this, however; it can utilize psychological or psychoanalytic theories to analyse *characters* in a novel, or to analyse the *reader* in the light of his or her responses to it. Take Henry James's novella *The Turn of the Screw*. A succession of critics has attempted to psychoanalyse the governess in this tale, finding evidence of mental instability of a sort that leads her to hallucinate and to imagine the ghosts that she reports on to the reader.

The problem that this raises is similar to that about which I talked earlier: treating a literary character as if he or she were a real person. Can it make sense to talk about psychoanalysing a literary character? Does a literary character have an Unconscious? Of course an author can choose to create a character who behaves like a real person, even to the extent of having repressions or being mentally ill, but it seems rather dangerous to assume that an author can unconsciously create a character who has the same sort of psychic life as a real person. (Indeed, what sense does this sort of suggestion make?)

Peter Lamarque has taken the critic Meredith Anne Skura to task for writing that 'In *Jane Eyre* . . . the manifest story about Jane's progress to adulthood is reinforced by a barely concealed female Oedipal fantasy in which growing up means marrying daddy.' He responds:

> [I]t is crucially unclear what status this alleged fantasy is supposed to have as a latent reading. Is it a feature of Charlotte Brontë's unconscious mind? Or the reader's? Or Jane's? A latent unconscious thought must belong to some mind or other on psychoanalytic theory. The only way a fantasy can be viewed as a property of a *work* is either by treating the work as a behavioral manifestation of an unconscious state of mind or by seeing the fantasy as an abstracted theme identified through literary critical methods. If the former, the latent content is not *literary*; if the latter, it is not *psychoanalytic*.[4]

Lamarque's objections sum up what many have seen as the problematic elements in using psychoanalytic theory in literary analysis, although he draws a sharper division between the literary critical and the psychoanalytic than I would.

Reader-oriented approaches

If we were to indulge in a vast and only partly justified generalization we could say that criticism of the novel up to the 1930s was often highly *biographical* in tendency – talking about the author as much as if not more than his or her work. From about this time onwards *textual* and *sociological* approaches to the study of the novel became more dominant (often in opposition to each other); while in the 1980s a growing interest in the *reader* and in the *reading process* could be detected.

Part of this can be explained as a reaction against New Critical treatments of the text of a novel as an object which can be studied objectively. On the contrary, recent critics have pointed out, the reading

---

[4] Peter Lamarque, *Fictional Points of View*, pp. 196–7.

of a novel is a *process*: we experience a novel as a sequence of responses over a period of time.

Other critics have suggested that it may well be misleading to talk of 'the reader' as if all readers (and readings) were the same. Virginia Woolf called a collection of her essays *The Common Reader* (1925), taking the phrase from Samuel Johnson and suggesting a respect for the ordinary as against the academic or specialist reader. But recent critics have started to talk of different conceptualizations of the reader: the implied reader, the original reader, the empirical reader.

The *implied reader* is that reader suggested by the text itself, most obviously when a narrator addresses comments directly to a reader but not only by such means. (Remember, however, my earlier warning that reader – implied or otherwise – and narratee are not necessarily the same.) When Jane Austen's narrator builds up an intimate relationship with her reader she does it by making assumptions about the reader: his or her values, interests, understanding. We can say, then, that we can almost build up an 'identikit' picture of the reader of a Jane Austen novel – the reader that we actually become when we read the novel in the way it was intended to be read. Thus, although two very different people may sit down to read *Emma*, in the process of reading that novel they may to some extent become very similar readers as they turn in to the sort of reader that the novel implies it expects. (Think how we respond to what people expect us to be when they talk to us.) Of course the gap between what we are and what the novel seems to want us to become may be so wide that we rebel at the pressure to conform. In such a case we may give the novel what students of the mass media have termed an *oppositional reading* – a reading that asserts values, interests and understanding contrary to those that we feel are implied in the work.

The *original reader* is obviously as much of a fictive construct as is that of the 'first-night audience' for a play. What it involves is an attempt to understand the novel in its historical context by asking what a sensitive, well-informed and intelligent reader would have made of the novel when first it was published. Note that this is not to say that an original reader is the same as an *ideal reader*; it may well be that later readers can see more in a novel than could have been seen on its first appearance. Many critics, without actually using the term, posit an ideal reader in their discussions of a novel: the reader who will respond in a way that maximizes what the novel can offer.

I think that the problem with such an assumption is that no single reading of a novel can combine all of that which a novel can offer a reader. As I have already pointed out, we get different things from a first and a second reading of the same novel and it is hard to see, logically, how these could be combined.

Moreover readers from different backgrounds can hardly be expected to read the same novel in exactly the same way. When we turn to *empirical readers* we have to assume, I think, that an accountant will read *Sons and Lovers* differently from the way in which a coal-miner will read it. What exactly the nature of this difference will be it is hard to state precisely, and it is possible to make a case for its being a matter of details rather than of essentials. It is worth spending some time to analyse yourself as a reader, however. To what extent are you likely to be influenced – even biased – in certain directions because of your background? Are there gaps in your knowledge or experience which may affect your understanding of this particular novel negatively? Can these be rectified in any way?

Feminist approaches

Just as feminists in society at large have encouraged us to look afresh at many aspects of our culture and history – down to quite small details of personal speech and behaviour – so too they have been extremely successful in getting us to look at literary works in new and often revealing ways. Over sixty-five years ago Virginia Woolf's *A Room of One's Own* (1929) startled readers by claiming that they lived in a patriarchy and that this fact conditioned the ways in which novels were written (or not written) and read. In more recent years a growing number of feminist critics have offered challenging accounts of the novel in general and of particular novels. Feminist critics have argued that not only have women had to overcome severe difficulties to become writers, but that once they have produced novels these have consistently been read in negative ways by male readers. Thus once a woman had managed to become literate (no easy matter in the past) she had to overcome all sorts of male prejudices in the reception accorded to her work. As Virginia Woolf expresses it:

> This is an important book, the critic assumes, because it deals with war. This is an insignificant book because it deals with the feelings of women in a drawing-room.[5]

---

[5] Virginia Woolf, *A Room of One's Own* (London, Hogarth Press, repr. 1967), p. 111.

Moreover, as Woolf points out, it may be difficult for a woman to write about such things as war because of the domestic role to which she has been confined, and she declares that a woman could not have written *War and Peace*.

Feminist critics have also done much to show the ways in which male views of reality have dominated much fiction – especially, of course, that by men, and especially their views of women. Women are typically portrayed *in relation to men*, and are often seen in certain stereotyped ways – as passive, hysterical, emotional, 'bitch' or 'goddess'. Thus it seems fair to say that it is only as a result of the efforts of feminist critics in recent years that the portrayal of women in D.H. Lawrence's major novels has been questioned and criticized, and that many other authors have been looked at with new eyes.

One of the works which introduced the new wave of feminist literary criticism was Kate Millett's *Sexual Politics*, first published in 1970. Millett's book is divided into three connected sections: 'Sexual Politics', giving a theoretical basis for what follows, a section on the historical background detailing the movements towards female emancipation and those seeking to reverse their effects, and finally a section entitled 'The Literary Reflection'. In this final section Millett looks at the work of D.H. Lawrence, Henry Miller, Norman Mailer, and Jean Genet. In the writing of all of these she detects what we now familiarly refer to as patriarchal ideas – frequently linked to portrayals that debase, ridicule and humiliate women. The following comments on Lawrence's *Lady Chatterley's Lover* are representative:

> In *Lady Chatterley*, as throughout his final period, Lawrence uses the word 'sexual' and 'phallic' interchangeably, so that the celebration of sexual passion for which the book is so renowned is largely a celebration of the penis of Oliver Mellors, gamekeeper and social prophet. While insisting his mission is the noble and necessary task of freeing sexual behavior of perverse inhibition, purging the fiction which describes it of prurient or prudish euphemism, Lawrence is really the evangelist of quite another cause – 'phallic consciousness'. This is far less a matter of 'the resurrection of the body', 'natural love', or other slogans under which it has been advertised, than the transformation of masculine ascendancy into a mystical religion, international, possibly institutionalized.[6]

---

[6] Kate Millett, *Sexual Politics* (London, Sphere Books, 1971), p. 238.

Millett's book is a fine piece of polemic, and one which jolted both men and women into reconsidering various canonical authors from previously unconsidered perspectives.

If this can be taken as exemplary of the anti-patriarchal thrust of the first stage of the feminist revival in literary criticism, feminist critics of literature in general and the novel in particular have also subjected writing by and for women (both canonical and popular) to analyses which combine familiar techniques with new and unfamiliar values.

A variety of feminist criticism which believes 'gendering' to take place at a more fundamental level concerns itself with what, using a French term, is termed *écriture féminine* – 'feminine writing'. Theorists of *écriture féminine* believe that men and women use language in fundamentally different ways, and that the task of the feminist critic is to give women writers the self-confidence to write as their femaleness requires, and not to ape the writing habits and style of men. As the term suggests, the concept in its modern form has originated in France, especially in the writing of Hélène Cixous. It has, however, antecedents and parallels outside France, and it is interesting in this context to see how many of the recent formulations about 'writing and the body' have parallels in Virginia Woolf's essay 'Professions for Women'.

A further task undertaken by feminist critics has been that of the 'rediscovery' or reinstatement of a number of (mainly women) writers whose works have been undervalued or forgotten. A number of publishing imprints devoted to encouraging the publication of works by women – both new authors and authors from the past – have been established, and the present high reputations of writers such as Jean Rhys, Doris Lessing, Tillie Olsen and others owe much (if not most) to the efforts of feminist critics.

Structuralist, post-structuralist and deconstructive approaches

1 *Structuralism*

Structuralist critics are, to start with a generalization, interested more in what makes meaning possible than in the different meanings that can be found – whether in the novel, in complete cultures (Structural Anthropology), or in a range of more specific cultural activities. We can use the analogy of the distinction within Linguistics between grammar and pragmatics. Whereas pragmatics focuses on particular utterances, the grammarian is concerned more with how meanings can be generated.

As this suggests, structuralism has much in common with certain approaches to the study of language, and is indeed characterized by a general belief that the structural study of language offers a paradigm (or model) for the study of other systems of meaning – including literature and the novel. This relationship is most clear in the work of a theorist mentioned earlier in this book – the Russian folklorist Vladimir Propp. Propp was interested not in the particular folk-tale, but in the system which allowed any given folk-tale to generate meaning. What appeared to be a surface difference between different folk-tales – one about a Princess, one about a miraculous goose – could be argued to involve an identity at a more fundamental level (both tales are to do with a pattern involving challenge, defeat, and final victory, for example). Propp's 'functions' were like grammatical functions: they could be actualized through a range of different examples just as the grammatical function of 'noun' can be actualized with 'dog', 'pig', or 'magistrate' in sentences with the same grammatical structure.

Structuralism in its purer form focuses far less on the detail of individual works than on the manner whereby they exemplify a set of possibilities open to the genre as a whole. Like the variant of Linguistics upon which they are based, literary critical structuralisms tend to be ahistorical, concentrating upon systems as they operate at a given moment of time rather than on their modification over time. In certain ways structuralist approaches resemble formalist approaches: they pay little or no attention to extrinsic factors, and consider either 'literature' or a genre such as the novel as to be a self-enclosed system that can be studied in terms of itself. Thus just as the New Critics argued against 'The Intentional Fallacy' of granting extra-textual statements of intent any prioritized weight in interpreting that text, so too structuralist critics have followed Roland Barthes in agreeing to announce 'The Death of the Author' so far as the interpretation of his or her text is concerned.

One objection to standard structuralist approaches has been that because of their ahistoricity they have difficulty in coming to terms with the ways in which novelists are able creatively to expand the realm of the possible. That system of possibilities that is the novel is not a static one; it is arguably expanded both by the importation of new possibilities from the world outside (the influence of new technologies, new developments in human culture and organization), and by formal experiments that may seem to break rules (and do), but which simultaneously change those rules. On the other hand, there is no doubt that narrative theory or narratology owes an enormous debt to structuralism, for it was structuralism's ability to look at narrative in general, as an abstract system which presented authors with a set of finite choices, that initiated the development of modern narrative theory.

2 Post-structuralism and deconstruction

For the purpose of a brief survey such as this, post-structuralism and deconstruction will be treated as synonymous. However, certain usages distinguish between the two terms in important ways and make post-structuralism a more general position of which deconstruction represents one possible variant.

Again, to simplify, we can say that post-structuralism extends and exaggerates structuralism, by taking the structuralist assumption that no element of a system of meaning has significance in and for itself but only as part of the system, a large step further. This involves seeing the generation of all meaning as a complete system, such that there is no bedrock, nothing firm upon which subsequent meanings can be constructed. Jacques Derrida, the French high-priest of deconstruction, calls such a desired point of stability in the swirl of meanings generated by the play between the different components of the system, a *transcendental signified*, in other words, an element that has meaning in and for itself. Thus when Derrida, famously (or infamously) declared that there was nothing outside of the text, he was arguing that there was no fixed, extra-textual point by reference to which the meaning of a text could be plotted and fixed.

Deconstructionists are thus committed to the unweaving of apparently firm and solid interpretations through a display of the way in which such interpretations rely upon such a transcendental signified. Examples of such transcendental signifieds in the criticism of the novel would be such things as 'the intended meaning of the author'; 'how contemporary readers read the work'; 'what the work means to a competent reader'; and so on. In practice, deconstructionist readings very often involve dispensing with such traditional critical points of reference as these, and responding to the work in a way that is unconstrained by them.

Postcolonial theory

The field of postcolonial theory is fashionable, fast-growing, and riven with debates. At its best it has confronted readers and critics with the hidden or unstated implications of colonial and imperialist relations between lands and people for the reading and interpretation of literary works. Thus my earlier discussion of the way in which Jean Rhys's *Wide*

Sargasso Sea interrogates Charlotte Brontë's *Jane Eyre* owes much to the new perspectives brought to literary and cultural criticism by postcolonialism. A key text is Edward Said's *Orientalism* (1978), which attempted to describe how western domination of subject lands required not just physical force but also a discourse, a system of ideas by which European culture was able to manage and even to produce 'the Orient'.

Following Said's example many writers attempted to use the vantage point of a period after colonialism ('post' colonialism) to interrogate literature written both during and after the classic forms of colonialism, and also to draw attention to the silenced voices of those who wrote from the standpoint of the oppressed.

This much is unproblematic – but much else is not. Earlier in this book I made reference to Ngũgĩ Wa Thiong'o's dedication of his novel *Devil on the Cross* 'To all Kenyans struggling against the neo-colonial stage of imperialism' (p. 51). If there are such things as neo-colonialism and imperialism, isn't the 'post' in postcolonialism a little optimistic? Does it refer to a historical or merely a mental and ideological super-session of colonialism? And if Australia and the United States are just as much postcolonial countries as are Kenya and India, then is not the term in danger of becoming so all-embracing that it loses any effective specificity? Many have pointed out that the term has become very popular precisely because few objections to it are raised in the academy: it avoids sensitive words such as 'imperialism' and it raises the topic of colonialism only to suggest that all that is safely in the past.[7]

For all this, postcolonialism has generated a number of extremely valuable studies and readings, and it represents a corrective to too unthinking an imposition of privileged viewpoints on the interpretation of literary works which touch on and were directly or indirectly supported by the under- or unprivileged.

---

[7] These and other issues are discussed in the following studies, especially in their Introductions: Padmini Mongia (ed.), *Contemporary Postcolonial Theory: A Reader* (London, Arnold, 1996); Gail Fincham and Myrtle Hooper (eds), *Under Postcolonial Eyes: Joseph Conrad After Empire* (Rondebosch, University of Cape Town Press, 1996).

Further Reading

Always try to ensure that you are not reading a novel in an inferior text. Many quite widely circulated paperback works have less than ideal texts, and it is easy to read a novel such as *Wuthering Heights* in a modern edition but in a form other than that which the author intended. The Penguin and World's Classics (Oxford University Press) texts maintain a high standard of textual reliability in the main, and the latter series includes critical introductions and notes as standard features. The Norton Critical Editions are the ideal, with excellent texts and a selection of important critical material in each volume. Also recommended is the 'Case Studies in Contemporary Criticism' series (Bedford Books/St Martin's Press; Macmillan Press in the UK). Volumes in this series also include reliable texts and a selection of scholarly and critical material.

Teachers and lecturers do not much like those 'Notes' that are little more than extended cribs, with plot summaries carefully provided for those who cannot be bothered to read the texts themselves. In some of these the critical comments are simplistic and unreliable. The best of this sort of series are the 'York Notes' published by Longman, and the Penguin Critical Studies. Volumes in both series are all written by very reputable scholars, and can be relied upon.

Other useful and reliable critical series are the Macmillan Casebooks and 'Critics Debate' texts, the Prentice-Hall Twentieth-Century Views and Twentieth-Century Interpretations, and the 'Open Guides to Literature' published by the Open University Press.

In addition, the following individual works can all be recommended. Note that I have not included here works to which I have made substantial reference in the body of the book.

Marguerite Alexander, *Themes and Strategies in Postmodernist British and American Fiction* (London, Edward Arnold, 1989).
> If you are curious about the form that postmodernism in fiction takes, and wish to read a sensible discussion of the work of such writers as Faulkner, Beckett, Pynchen and Vonnegut, then this is the place you are recommended to start.

Wayne C. Booth, *The Rhetoric of Fiction* (Chicago, University of Chicago Press, 1961).

A study that really initiated the new wave of interest in narrative technique. Booth is a stimulating and accessible critic, and his book has been enormously influential. The second edition contains an interesting 'Afterword'.

Malcolm Bradbury (ed.), *The Novel Today: Contemporary Writers on Modern Fiction* (Glasgow, Fontana, revised edition 1990).

A very useful collection of essays by practising novelists writing about their own, and others', fiction. Essential for any student interested in modernism and postmodernism in the novel. The 1990 edition includes a new introduction, a new piece by Italo Calvino, and an interview with Milan Kundera by Ian McEwan.

Anne Cranny-Francis, *Feminist Fiction: Feminist Uses of Generic Fiction* (Oxford, Polity Press, 1990).

This is an intelligent study of what is claimed to be a range of feminist genres: feminist science fiction, fantasy, utopias, detective fiction, and romance. If you are interested in what a feminist investigation into these areas, and into the work of important recent writers such as Angela Carter, Marge Piercy, Valerie Miner, and others looks like, then this is the place to start. The book also contains a useful concluding chapter on 'Gender and Genre'.

Monika Fludernik, *The Fictions of Language and the Languages of Fiction* (London, Routledge, 1993).

Along with Ann Banfield's *Unspeakable Sentences* (see p. 94), on which it builds but with which it takes issue on a number of points, probably the most exhaustive and demanding study of Free Indirect Discourse available. Not recommended as an introductory text, but very rewarding for the reader prepared to grapple with its difficulties.

Alan Warren Friedman, *Fictional Death and the Modernist Enterprise* (Cambridge, Cambridge University Press, 1995).

Starting from the premise that different cultures react to death in varied ways, Friedman notes how modernist novelists responded to and portrayed death in very dissimilar ways from their Victorian predecessors. Contains valuable chapters on Forster, Woolf, Greene, and Durrell, in addition to its more extended socio-cultural investigation into twentieth-century attitudes to death and dying.

Robin Gilmour, *The Novel in the Victorian Age* (London, Edward Arnold, 1986).

There are many who claim that the Victorian age is the high point of achievement of the English novel. Whether or not one agrees, this is one of the best introductions to the Victorian English novel, intelligently blending information about the age with detailed commentary on individual novelists and novels.

Geoffrey N. Leech and Michael H. Short, *Style in Fiction: A Linguistic Introduction to English Fictional Prose* (London, Longman, 1981).

Those who have tried to read accounts of prose fiction by Linguisticians and who have determined not to try again should relent and read this book. The authors yield nothing to literary specialists in their sensitivity to fictional texts and passages, and are able to present a conception and analysis of style which any student of the novel will find invaluable. Well provided with sensible and well-illustrated examples.

David Lodge, *The Art of Fiction* (Harmondsworth, Penguin, 1992).

A collection of short articles originally published in *The Independent on Sunday*, offering both accessible and authoritative accounts of fifty different terms and concepts ranging from 'Beginning' to 'Ending' and taking in others such as suspense, stream of consciousness, defamiliarization, repetition, showing and telling, symbolism, allegory, epiphany, metafiction, aporia, and the Uncanny. All are illustrated with extracts from either classic or modern texts. A must for the student of the novel or of narrative.

Brian McHale, *Postmodernist Fiction* (London, Methuen, 1987; now published by Routledge).

McHale's study is extremely wide-ranging, including authors and texts from several continents, and theoretically penetrating and insightful. McHale is very well versed in modern narrative theory, but he brings other theoretical perspectives to bear on his analyses.

Adam Zachary Newton, *Narrative Ethics* (Cambridge, Mass., Harvard University Press, 1995).

Newton directs our attention to the ethical implications which telling stories and creating characters have for both author and reader. A very useful corrective to purely 'technical' studies of narrative technique.

Contains important readings of work by (among others) Conrad, Anderson, James, Dickens, and Julian Barnes.

Lynne Pearce, *Reading Dialogics* (London, Arnold, 1994).
For those interested in a recent application of Mikhail Bakhtin's theories of dialogism to a range of texts, this book is recommended. Pearce includes both readings of individual works (including *Wuthering Heights*, *The Waves*, *Sexing the Cherry* and *Beloved*) and an extended theoretical discussion of dialogue and the dialogic.

Gill Plain, *Women's Fiction of the Second World War: Gender, Power and Resistance* (Edinburgh, Edinburgh University Press, 1996).
An interesting feminist study of the different ways in which women writers respond to that 'crisis of patriarchy' which is war. Includes commentary on the work of Dorothy L. Sayers, Stevie Smith, Virginia Woolf, Naomi Mitchison and Elizabeth Bowen. A useful bringing together of two contexts: that of socio-historical crisis, and that of the community of women novelists.

Martin Seymour-Smith, *Novels and Novelists: A Guide to the World of Fiction* (New York, St Martin's Press, 1980).
Although this has the outward appearance of a coffee-table book, and its 'star-ratings' of major novels may strike a discordant note, it actually contains some excellent sections written by highly respected academics. Seymour-Smith himself provides a splendid opening chapter on the origins and development of the novel, J.A. Sutherland provides chapters on approaches to the novel and the novel and the book trade, Gamini Salgado writes on the novelist at work, Michael Mason on fiction and illustration, and David Pirie contributes a thought-provoking chapter on the novel and the cinema.

Dale Spender, *Mothers of the Novel: 100 Good Women Writers Before Jane Austen* (London, Pandora Press, 1986).
The title is relatively self-explanatory: this is an attempt to reclaim the work of many 'forgotten' female contributors to the development of a genre which owes an enormous debt to the female sex. For those suspicious of the canon and of its likely patriarchal biases this is a good place to start.

Mark Spilka (ed.), *Towards a Poetics of Fiction* (London, Indiana University Press, 1977).

This book contains essays from the journal *Novel* from the period 1967–1976, and includes many fine theoretical and interpretative pieces by well-known writers. Wayne C. Booth and Ian Watt comment, respectively, on their influential books *The Rhetoric of Fiction* and *The Rise of the Novel*. There are also studies of the criticism of Georg Lukács, Wayne C. Booth, and F.R. Leavis, and an essay on the contributions of formalism and structuralism to the theory of fiction. This is a major source for ideas about the novel, not simple, but well worth study.

Leona Toker, *Eloquent Reticence: Withholding Information in Fictional Narrative* (Lexington, University Press of Kentucky, 1993).

A must for anyone interested in how the creative powers of readers contribute to the experience of reading novels. Toker takes a number of examples in which the reader has to 'fill in' missing or delayed information. There are excellent chapters on *The Sound and the Fury*, *Nostromo*, *Bleak House*, *Emma*, *Tom Jones*, *A Passage to India*, and *Absalom, Absalom!*.

J.M.S. Tomkins, *The Popular Novel in England 1770–1800* (London, Methuen, 1961; first published London, Constable, 1932).

A book that has dated very little, and essential reading for those interested either in the importance to the novel of women as writers and readers or in the whole complex issue of the novel and 'the popular'. Particularly interesting on such matters as the theme of incest in early modern novels and the symbolic importance of the brother-sister relationship to early women novelists.

Michael J. Toolan, *Narrative: A Critical Linguistic Introduction* (London, Routledge, 1988).

An accessible introduction to narrative theory which includes a concern not just with literary narratives but also with oral narratives and narratives in newspapers, law courts and elsewhere. Although Toolan's emphasis is a linguistic one, he is good on the ideological and political ramifications of narrative choices.

Index